Churchill Babington

An introductory lecture on archaeology delivered before the University of Cambridge

Churchill Babington

An introductory lecture on archaeology delivered before the University of Cambridge

ISBN/EAN: 9783742803467

Manufactured in Europe, USA, Canada, Australia, Japa

Cover: Foto ©berggeist007 / pixelio.de

Manufactured and distributed by brebook publishing software (www.brebook.com)

Churchill Babington

An introductory lecture on archaeology delivered before the University of Cambridge

INTRODUCTORY LECTURE

ON

ARCHÆOLOGY.

Cambridge:
PRINTED BY C. J. CLAY, M.A.
AT THE UNIVERSITY PRESS.

AN

INTRODUCTORY LECTURE

ON

ARCHÆOLOGY

Delivered before the University of Cambridge.

BY

CHURCHILL BABINGTON, B.D., F.L.S.

DISNEY PROFESSOR OF ARCHÆOLOGY, SENIOR FELLOW OF ST JOHN'S COLLEGE,
MEMBER OF THE ROYAL SOCIETY OF LITERATURE, OF THE NUMISMATIC
AND SYRO-EGYPTIAN SOCIETIES, HONORARY MEMBER OF THE
HISTORICO-THEOLOGICAL SOCIETY OF LEIPSIC, AND OF
THE ARCHÆOLOGICAL INSTITUTE OF ROME.

CAMBRIDGE:
DEIGHTON, BELL, AND CO.
LONDON: BELL AND DALDY.
1865.

PREFACE.

The following Lecture was divided in the delivery into two parts; illustrative specimens being exhibited after the conclusion of the delivery of each portion. It has been suggested that I should add in the form of notes a few books which may prove useful to the students of particular branches of Archæology; my best thanks are due to the Rev. T. G. Bonney and the Rev. W. G. Searle for their kind and valuable assistance in drawing up certain of the lists. For ancient art and archæology K. O. Müller's Manual, so often referred to, will in general sufficiently indicate the bibliography, and it is only in a few departments, in numismatics more especially, that it has been deemed necessary to add anything to his references. M. Labarte's Handbook, from which a great part of the concluding portion of this lecture is derived, will do the same thing, though in a far less complete manner, for medieval art.

CONTENTS.

PLAN of the Lecture, pp. 1—3.

Archæology defined, and the principal kinds of archæological monuments specified, pp. 3—6.

Nature of the Disney Professorship of Archæology explained; its comprehensive character; the advantages of this, pp. 6—13.

Sketch of the existing remains of Antiquity among different nations, beginning with primeval man, pp. 13—21. The Egyptians, pp. 21—26. The Babylonians, pp. 26, 27. The Assyrians, pp. 27, 28. The Persians, pp. 28, 29. The Jews, pp. 29—31. The Phœnicians, pp. 31, 32. The Lycians, pp. 32, 33. The Greeks, pp. 33—41. The Etruscans, p. 41. The Romans, pp. 42—46. The Celts, pp. 43, 44. The Byzantine empire and the European nations during the middle ages, pp. 46—61. Recapitulation, pp. 61, 62.

Qualifications necessary for an archæologist. He must be a collector of facts and objects, and be able to reason on them. He must also be a man of learning. Exact scholarship, an appreciation of art, and a knowledge of natural history often useful or necessary for the archæologist, pp. 63—68.

Pleasures and advantages which result from archæology. It illustrates and is illustrated by ancient literature. Modern art aided by archæology. Archæology deserving of cultivation for its own sake, as an ennobling and delightful pursuit, pp. 68—74.

INTRODUCTORY LECTURE

ON

ARCHÆOLOGY.

Following the example of my distinguished predecessor in the Disney Professorship of Archæology, I open my first Course of Lectures with an introductory Lecture on Archæology itself, so far as the very limited time for preparation has allowed me to attempt one.

I cannot indeed conceal from myself, and still less can I conceal from you, that no introductory Lecture which I could give, even if I were to take my own time in writing it, would bear any comparison with the compositions of his elegant and learned pen. It certainly does not proceed from flattery, and I hope not from an undue partiality of friendship to say of him, that in his power of grasping a complicated subject, of presenting it in a clear light, of illustrating it with varied learning, and of expressing himself in relation thereto in appropriate language, I have rarely seen his equal. To how great a disadvantage then must I necessarily appear, when I have had only six 'weeks' time in which to get ready this as well as five other Lectures, and have been moreover compelled to devote a considerable part even of that short time to other and not less important duties. A great unwillingness however that the Academical year should pass over without any Archæological Lectures being delivered by the Disney Professor, has induced me to make the attempt more quickly than would under other circumstances have been desirable or even justifiable; and I venture to hope that when allowance is made for the exigency of the case, I shall find in you, who have honoured this Lecture by your presence, a clement and even an indulgent audience.

In an introductory Lecture which deals with generalities, it is hardly to be expected that I either can say or ought to try to say much which is absolutely new to any of my hearers; and I shall not affect to say anything peculiarly striking, but shall rather attempt to bring before you in a plain way a view of the subject, which aims at being concise and comprehensive; and in connexion therewith respectfully to submit a few observations which have relation to other Academical studies, as well as to the character of this particular Professorship.

What I propose then to do is this, first to explain what Archæology is; next to put in a clear light what the character of this Professorship is;

after that to attempt a general sketch of the existing remains of Antiquity; then to point out the qualifications necessary or desirable for an archæologist; and in conclusion, to indicate the pleasure and advantage which flow from his pursuits.

The field of Archæology is vast, and almost boundless; the eye, even the most experienced eye, can hardly take in the whole prospect; and those who have most assiduously laboured in its exploration will be most ready to admit, that there are portions, and those large portions, which are to them either almost or altogether unknown.

For what is Archæology? It is, I conceive, the science of teaching history by its monuments[1], of whatever character those monuments may be. When I say history, I use the word not in the limited sense of the history of dynasties or of governments. Archæology does indeed concern itself with these, and splendidly does it illustrate and illuminate them; but it also concerns itself with every kind of monument of man which the ravages of time have spared.

[1] Perhaps it would be more correct to say 'by its *contemporary sensible* monuments,' so as to exclude later copies of ancient writings, or the *monumenta litterarum*, which fall more especially to the province of the scholar. A MS. of Aristotle of the thirteenth century is an archæological monument of that century only; it is a literary monument of the fourth century B.C. But a Greek epigram or epitaph which occurs on a sepulchral monument of the same or any other century B.C. is an archæological as well as a literary monument of that century.

Archæology concerns itself with the domestic and the social, as well as with the religious, the commercial, and the political life of all nations and of all tribes in the ages that have passed away. All that men in ancient times have made, and left behind them, is the farrago of our study. The archæologist will consequently make observations and speculations on the sites of ancient cities where men have dwelt; on their walls and buildings, sacred and profane; on their altars and their market-places; on their subterranean constructions, whether sepulchres, treasuries, or drains. He will trace the roads and the fosses along which men of the old world moved, and on which men often still move; he will explore the routes of armies and the camps where they have pitched, and will prowl about the barrows in which they sleep;

> Exesa inveniet scabra robigine pila,
> Grandiaque effossis mirabitur ossa sepulchris.

He will also collect and classify every kind of object, which man has made for use or for ornament in his own home, or in the city; in the fields, or on the water. He will arrange the weapons of offence and defence according to their material and age; whether of stone, of bronze, of iron, or of steel; among which some are so rude that a practised eye alone distinguishes them from the broken flint stones lying in the field, others again so elaborate as to rank among the most beautiful productions both of classical and medieval art; he will not disdain to preserve the bricks and the tiles, which have once formed parts of Asiatic cities or of Roman farms; he will excavate the villas of the ancients; unearth their mosaic pavements; clean their lamps and candelabra; he will mend or restore their broken crockery, and glass; he will even penetrate into the lady's chamber, turn over her toilet, admire her brooches and her bracelets, examine her mirrors and her pins; and all this he will do in addition to studying the nobler works of ancient art, such as engraved gems and medallions; works chased, carved and embossed in the precious metals and in ivory; frescoes and vase-paintings; bronzes and statues. He will, likewise, familiarise himself with the alphabets of the ancient nations, and exercise his ingenuity in deciphering their written records, both public and private; whether these be contained in inscriptions on stones or metal plates, or in papyrus-rolls, or parchment books; or be scratched on walls or on statues; or be painted on vases; or, in fine, surround the device of a coin.

I have now mentioned some of the principal objects of archæology, which, as I have said, embraces within its range all the monuments of the history and life of man in times past. And this it does, beginning with the remains of primeval man, which stretch far beyond the records of all literary history, and descending along the stream of time till it approaches, but does not quite reach time actually present. No sharp line of demarcation separates the past from the present; you may say that classical archæology terminates with the overthrow of the Western Empire; you may conceive that medieval archæology ceases with the reign of Henry the Seventh; but, be this as it may, in a very few generations the objects of use or of ornament to us will become the objects of research to the archæologist; and, 1 may add, may be the subjects of lectures to my successors.

For the founder of this Professorship, whose memory is never to be named without honour, and the University which accepted it, together with his valuable collection of ancient sculptures, undoubtedly intended that any kind or class of antiquities whatever might fitly form the theme of the Professor's discourse. I say this, because a misconception has undoubtedly prevailed on this subject, from which even my learned predecessor himself was not free. "Every nation of course," says he, "has its own peculiar archæology. Whether civilized or uncivilized, whether of historic fame or of obscure barbarism, Judæa, Assyria, and Egypt; Greece and Rome; India, China, and Mexico; Denmark, Germany, Britain, and the other nations of modern Europe, all have their archæology. The field of inquiry," he continues, "is boundless, and in the multitude of objects presenting themselves the enquirer is bewildered. It has been wisely provided therefore by the founder of this Professorship, that we shall direct our attention more immediately to one particular class of Antiquities, and that the noblest and most important of them all, I mean the Antiquities of Greece and Rome[1]." Very probably such may have been Mr Disney's original intention; and if so, this will easily explain and abundantly pardon the error of my accomplished friend; but the actual words of the declaration and agreement between Mr Disney and the University, which is of course the only document of binding force, are as follows: "That it shall be the duty of the Professor to deliver in the course of each academical year, at such days and hours as the Vice-Chancellor shall appoint, six lectures at least on the subject of Classical, Mediæval and other Antiquities, the Fine Arts and all matters and things connected therewith." Whether he would have acted wisely or not wisely in limiting the field to classical archæology, he has in point of fact not thus limited it. And, upon the whole, I must confess, I am glad that he has imposed no limitation. For while there are but few who would deny that many of the very choicest relics of ancient art and of ancient history are to be sought for in the Greek and Roman saloons and cabinets of the museums of Europe, yet it must at the same time be admitted that there are other branches of archæology, which are far too

[1] Marsden's *Introd. Lect.* p. 5. Cambr. 1852.

important to be neglected, and which have an interest, and often a very high interest, of their own.

Let it be confessed, that the archæology of Greece has in many respects the pre-eminence over every other. "It is to Greece that the whole civilized world looks up," says Canon Marsden, "as its teacher in literature and in art; and it is to her productions that we refer as the standard of all that is beautiful, noble, and excellent. Greece excelled in all that she put her hand to. Her sons were poets and orators and historians; they were architects and sculptors and painters. The scantiest gleanings of her soil are superior to that which constitutes the pride and boast of others. Scarcely a fragment is picked up from the majestic ruin, which does not induce a train of thought upon the marvellous grace and beauty which must have characterized the whole!

> Qualo te dicat tamen
> Antehac fuisse, tales cum sint relliquiæ."

These eloquent and fervid words proceed from a passionate admirer of Hellenic art, and a most successful cultivator of its archæology. Nor do I dare to say that the praise is exaggerated. But at the same time, viewed in other aspects, the archæology of our own country has even greater interest and importance for us. What man is there, in whose breast glows a spark of patriotism, who does not view the monuments of his country which are everywhere spread around him, (in this place above most places,) which connect the present with the remote past, and with many and diverse ages of that past by a thousand reminiscences, with feelings deeper and nobler than any exotic remains of antiquity, how charming soever, could either foment or engender? This love of national antiquities, seated in a healthy patriotic feeling, has place in the speech of an apostle himself: "Men and brethren, let me freely speak unto you of the patriarch David, that he is both dead and buried; and his sepulchre is with us unto this day." The same feeling prompted Wordsworth thus to express himself in reference to our ancient colleges and their former occupants:

> I could not always lightly pass
> Through the same gateways, sleep where they had slept,
> Wake where they waked; I could not always print
> Ground where the grass had yielded to the steps
> Of generations of illustrious men,
> Unmoved......
>   Their several memories here
> Put on a lowly and a touching grace
> Of more distinct humanity.

And not only the buildings, but the other archæological monuments of the University (for so I think I may be permitted to call the pictures and the busts, and the statues, and the tombs, which are the glories of our chapels, our libraries and our halls) teach the same great lessons. They raise up again our own worthies before our very eyes, calling on us to strive to walk as they walked, dead though they be and buried; for their effigies and their sepulchres are 'with us to this day.' I must repeat, then, that I am glad that the Disney Professor is not obliged to confine himself to classical archæology, sorry as I should be if he were wholly unable to give lectures on one or more branches of that most interesting department, which has moreover a special connexion with the classical studies of the University. It is manifest that the University intended the Professor to consider no kind of human antiquities as alien from him; and I think this in itself a very great gain. For, if the truth must be confessed, antiquaries above most others have been guilty of the error of despising those branches of study which are not precisely their own. I forbear to adduce proofs of this, though I am not unprovided with them; and even although you would certainly be amused if I were to read them; classicists against gothicists; gothicists against classicists.

I could wish that the learned and meritorious writers on both sides had profited by the judicious remarks of Mr Willson, prefixed to Mr Pugin's *Specimens of Gothic Architecture in England.* "The respective beauties and conveniences proper to the Grecian orders in their pure state or as modified by the Romans and their successors in the Palladian school may be fully allowed, without a bigoted exclusion of the style we are accustomed to term Gothic. Nor ought its merits to be asserted to the disadvantage of the classic style. Each has its beauties, each has its proportions[1]." One of the most eminent Gothic architects, Mr George Gilbert Scott, expresses himself in a very similar spirit. "It may be asked, what influence do we expect that the present so called classic styles will exercise upon the result we are imagining, (i. e. the devolopement of the architecture of the future). Is the work of three centuries to be unfelt in the future developements, and are its monuments to remain among us in a state of isolation, exercising no influence upon future art? It would, I am convinced, be as unphilosophical to wish, as it would be unreasonable to expect this[2]." To turn from them to the classicists. "See how much Athens gains," says Prof. T. L. Donaldson, "upon the affections of every people, of every age, by her Architectural ruins. Not a traveller visits Greece whose chief purpose is not centred in the Acropolis of Minerva.... But in thus rendering the homage due to ancient Art it were unjust to pass without notice those sublime edifices due to the Genius of our Fathers. It is now unnecessary to enter upon the question, whether the first ideas of Gothic Architecture were the result of a casual combination of lines or a felicitous adaptation of form derived immediately

---

[1] P. xix. London, 1821.
[2] Scott's *Remarks on Secular and Domestic Architecture, present and future*, p. 272. London, 1857.

from Nature: But graceful proportion, solemnity of effect, variety of plan, playfulness of outline and the profoundest elements of knowledge of construction place these edifices on a par with any of ancient times. Less pure in conception and detail, they excel in extent of plan and of disposition, and yield not in the mysterious effect produced on the feelings of the worshipper. The sculptured presence of the frowning Jove or the chryselephantine statue of Minerva were necessary to awe the Heathen into devotion. But the presence of the Godhead appears, not materially but spiritually, to pervade the whole atmosphere of one of our Gothic Cathedrals[1]." The Editor of *The Museum of Classical Antiquities*, well says, "As antiquity embraces all knowledge, so investigations into it must be distinct and various. Each antiquary labours for his own particular object, and each severally assists the other[2]." It should be borne in mind moreover that archæological remains of every kind and sort are really a part of human history; and if all parts of history deserve to be studied, as they most assuredly do, being parts, though not equally important parts, of the Epic unity of our race, it will follow even with mathematical precision that all monuments

[1] *Preliminary Discourse* pronounced before the University College of London, upon the commencement of a series of Lectures on Architecture, pp. 17—24. London, 1842.
[2] *Museum of Classical Antiquities*, Vol. L p. 1. London, 1851.

relating to all parts of that history must be worthy of study also.

I desire therefore to express in language as strong as may be consistent with propriety, my entire disapproval of pitting one branch of archæology against another, or indeed any study against another study. And on this very account I rejoice that the Disney Professor's field of choice is as wide as the world itself, so far as concerns its archæology. There is no country, there is no period about which he may not occupy himself, or on which he may not lecture, if he feel himself qualified to do so. He is in a manner bound by the tenure of his office to treat every branch of archæology with honourable respect; and this in itself may not be without a wholesome influence both upon his words and sentiments. I have been somewhat longer over this matter than I could have wished; but I thought it desirable that the position of the Disney Professor should be rightly understood; and I have also endeavoured to shew the real advantage of that position.

His field then is the world itself; but as this is so (and as I think rightly so) there is a very true and real danger lest he and his hearers should be mazed and bewildered at the contemplation of its magnitude. Yet in spite of that danger I will venture to invite you to follow the outlines of the great entirety of the relics of the ages that have for ever passed away. I say the outlines, and even this is almost too much, for I am compelled to shade some parts of the picture so obscurely, and to throw so much of other parts into the background, that even of the outlines I can distinctly present to you but a portion. Thus I will say little more of the archæology of the New World, than that there is one which reaches far beyond the period of Spanish conquest, comprising among many other things ruins of Mexican cities, exquisite monuments of bas-reliefs and other carvings in stone; I will not invite you into the far East of the Old World, to explore the long walls and Buddhist temples of the ancient and stationary civilisation of China, or to dwell upon the objects of its fictile and other arts; but leaving both this and all the adjacent countries of Thibet, Japan and even India without further notice, or with only passing allusions, *spatiis conclusus iniquis*, I will endeavour, so far as my very limited knowledge permits, the delineation of the most salient peculiarities of the various remains of the old world till the fall of the Roman Empire in the West, and then attempt to trace briefly the remains of successive medieval classes of antiquities, until we arrive at almost modern times. I can name but few objects under each division of the vast subject; but these will be selected so as to suggest as much as possible others of a kindred kind. In addressing myself to such an audience, I may, if anywhere, act upon the assumption, *Verbum sapienti sat est:* a single word may suggest a train of thought. If I cannot wholly escape the charge of tediousness, I must still be content: for I am firmly convinced after the most careful consideration that I can pursue no course which is equally profitable, though I might take many others which might be more amusing.

It would now appear probable that the earliest extant remains of human handicraft or skill have as yet been found, not on the banks of the Nile or the Euphrates, but in the drift and in the caverns of Western Europe. Only yesterday, as I may say, it has been found out that in a geological period when the reindeer was the denizen of Southern France, and when the climate was possibly arctic, there dwelt in the caverns of the Périgord a race of men, who were unacquainted with the use of metals, but who made flint and bone weapons and instruments; who lived by fishing and the chase, eating the flesh of the reindeer, the aurochs, the wild goat and the chamois; using their skins for clothes which they stitched with bone needles, and their bones for weapon handles, on which they have etched representations of the animals themselves. Specimens of these things were placed last year in the British Museum; and a full account of the discoveries in 1862 and 1863 may be seen in the *Revue Archéologique*. Some distinguished antiquaries consider that they are the earliest human remains in Western Europe. Various other discoveries in the same regions of late years have tended towards shewing that the time during which man has lived upon the earth is much greater than we had commonly supposed. The geological and archæological circumstances under which the flint implements were found at Abbeville, and St Acheul, near Amiens, in the valley of the Somme, left no doubt that they were anterior by many ages to the Roman Empire. They have a few points of similarity to those found in the caverns of the Périgord, and as they occur along with the remains of the Elephas antiquus and the hippopotamus, Sir Charles Lyell infers that both these animals coexisted with man; and perhaps on the whole we may consider them rather than those of the Périgord to be the earliest European remains of man, or of man at all. Similar weapons have been found in the drift in this country, in Suffolk, Bedfordshire, and elsewhere. At Brixham, near Torquay, a cavern was examined in 1858, covered with a floor of stalagmite, in which were imbedded bones of the reindeer and also an entire hind leg of the extinct cave-bear, every bone of which was in its proper place; the leg must consequently have been deposited there when the separate bones were held together by their ligaments. Below this floor was a mass of loam or bone-earth, varying from one to fifteen feet in thickness, and amongst it, and the gravel lying below it, were discovered about fifteen flint knives, recognised by practised archæologists as artificially formed, and among them one very perfect tool close to the leg of the bear. It thus becomes manifest that the extinct bear lived after the flint tools were made, or at any rate not earlier; so that man in this district was either the contemporary of the cave-bear, or (as would seem more probable) his predecessor. But shortness of time forbids me to do more than to indicate that in western Europe generally, as well as in Britain, we have an archæology beginning with the age of the extinct animals or quaternary geological epoch and connecting itself with the age of the Roman Empire, when the first literary notices of those countries, with slight exceptions, commence. The antiquaries and naturalists of Denmark conjointly (these indeed should always be united, having much in common; and I am happy in being able to say that a love of archæology has often been united with a love of natural science by members of this University, among whom the late and the present Professor of Botany may be quoted as examples)—these Danish archæologists and naturalists I say, have made out three distinct periods during this interval: the age of stone contemporary with the pine forests; the age of bronze commencing with the oak forests which lie over the pine in the peat; and the age of iron co-extensive with the beech forests which succeeded the oak, and which covered the country in the Roman times as they cover it now. The skulls belonging to the oldest or stone age resemble those of the modern Laplanders; those of the second and third are of a more elongated type.

The refuse-heaps along the shores of the islands of the Baltic, consisting of the remains of mollusks and vertebrated animals, mingled with stone weapons, prove the great antiquity of the age of stone; the oyster then flourished in places where, by reason of the exclusion of the ocean from the brackish Baltic, it does not now exist. None of the animals now extinct, however, occur in these Kjökkenmödding, as they are called, except the wild bull, the *Bos primigenius*, which was alive in Roman times; but the bones of the auk, now, in all probability, extinct in Europe, are frequent; also those of the capercailzie, now very rare in the southern districts of Scandinavia, though abundant in Norway, which would find abundant food in the buds of the pines growing in prehistoric times in the peat bogs. Similar refuse-heaps, left in Massachusetts and in Georgia by the North American Indians, are considered by Sir C. Lyell, who has seen them, to have been there for centuries before the white man arrived. They have also been found, I understand, very recently in Scotland in Caithness. The stone weapons have now been sharpened by rubbing, and are less rude and probably more recent than those of the drift of the Somme valley, or of the caverns of the Périgord. The only domestic animal belonging to the stone age, yet found in Scandinavia, is the dog; and even this appears to have been wanting in France. In the ages of bronze and iron various domestic animals existed;

but no cereal grains, as it would seem, in the whole of Scandinavia. Weapons and tools belonging to these three periods, as well as fragments of pottery and other articles, are very widely diffused over Europe, and have been met with in great abundance in our own country (in Ireland more especially), as well as near the Swiss-lake habitations, built on piles, to which attention has only been called since 1853. It is strange that all the Lake settlements of the bronze period are confined to West and Central Switzerland: in the more Eastern Lakes those of the stone period alone have been discovered.

Similar habitations of a Pæonian tribe dwelling in Lake Prasias, in modern Roumelia, are mentioned by Herodotus, and they may be compared, in some degree, with the Irish Lake-dwellings or Crannoges, *i.e.* artificial islands, and more especially with the stockaded islands, occurring in various parts of the country: and which are accompanied by the weapons and instruments and pottery of the three aforesaid periods. Even in England slight traces of similar dwellings have been found near Thetford, not accompanied by any antiquities, but by the bones of various animals, the goat, the pig, the red deer, and the extinct ox, the *Bos longifrons*, the skulls of which last were in almost all instances fractured by the butcher.

As to the chronology and duration of the three periods I shall say nothing, though not ignorant that some attempts have been made to determine them. They must have comprehended several thousand years, but how many seems at present extremely uncertain. I should perhaps say that Greek coins of Marseilles, which would probably be of the age of the Roman Republic, have been found in Switzerland in some few aquatic stations, and in tumuli among bronze and iron implements mixed. The cereals wanting in Scandinavia appear in Switzerland from the most remote period; and domestic animals, the ox, sheep, and goat, as well as the dog, even in the earliest stone-settlements. Among the ancient mounds of the valley of the Ohio, in North America, have been found (besides pottery and sculpture and various articles in silver and copper) stone weapons much resembling those discovered in France and other places in Europe. Before passing from these pre-historic remains, as they are badly called, to the historic, let me beg you to observe a striking illustration of the relation of archæology to history. Archæology is not the handmaid of history; she occupies a far higher position than that: archæology is, as I said at the outset, the science of teaching history by its monuments. Now for all western and northern Europe nearly the whole of its early history must be deduced, so far as it can be deduced at all, from the monuments themselves; for the so-called monuments of literature afford scanty aid, and for that reason our knowledge of these early ages is necessarily very incomplete. Doubtless, many a brave Hector and many a brave Agamemnon lived, fought, and died in the ages of stone and of bronze; but they are oppressed in eternal night, unwept and unknown, because no Scandinavian Homer has recorded their illustrious deeds. Still, we must be thankful for what we can get; and if archæological remains (on which not a letter of an alphabet is inscribed) cannot tell us everything, yet, at least, everything that we do know about these ages, or very nearly so, is deduced by archæology alone.

We must now take a few rapid glances at the remains of the great civilised nations of the ancient world. Mr Kenrick observes that the seats of its earliest civilisation extend across southern Asia in a chain, of which China forms the Eastern, and Egypt the Western extremity; Syria, Mesopotamia, Assyria, and India, are the intermediate links. In all these countries, when they become known to us, we find the people cultivating the soil, dwelling in cities, and practising the mechanical arts, while their neighbours lie in barbarism and ignorance. We cannot, he thinks, fix by direct historical evidence the transmission of this earliest civilisation from one country to another. But we may determine with which of them ancient history and archæology must begin. The monuments of Egypt surpass those of all the rest, as it would appear, by many centuries. None of the others exercised much influence on European civilisation till a later period, some exception being made for the Phœnician commerce; but the connection of

European with Egyptian civilisation is both direct and important. "From Egypt," he remarks, "it came to Greece, from Greece to Rome, from Rome to the remoter nations of the West, by whom it has been carried throughout the globe[1]." As regards its archæology, which is very peculiar and indeed in some respects unique, I must now say a few words. The present remains of Memphis, the earliest capital, said to have been founded by Athothis, the son of Menes, the first king of the first dynasty, are not great; but so late as the fourteenth century they were very considerable. Temples and gateways, colossal statues and colossal lions then existed, which are now no more. Whether any of them approached the date of the foundation it is useless to enquire. Now, the most remarkable relic is a colossal statue of Rameses II., which, when perfect, must have been about forty-three feet high. This monarch is of the XVIIIth dynasty, which embraces the most splendid and flourishing period of Egyptian history; and though much uncertainty still prevails for the early Egyptian chronology, it appears to be well made out and agreed that this dynasty began to reign about fifteen centuries before the Christian era. But the pyramids and tombs of Ghizeh, and of several other places at no great distance from Memphis, are of a much earlier date; and the great pyramid is securely referred

[1] *Ancient Egypt*, Vol. I. p. 3. London, 1850.

to a king of the fourth dynasty. "Probably at no place in the entire history of Egypt," says Mr Osburn, "do the lists and the Greek authors harmonize better with the historical notices on the monuments than at the commencement of this dynasty[1]." The system of hieroglyphic writing was the same (according to Mr Kenrick) in all its leading peculiarities, as it continued to the end of the monarchy. I regret to say that some eminent men have tried to throw discredit, and even ridicule, on the attempts which, I think, have been most laudably made with great patience, great acuteness, and great learning, to decipher and interpret the Egyptian and other ancient languages. Many of us, doubtless, have seen a piece of pleasantry in which *Heigh-diddle-diddle, The cat and the fiddle* is treated as an unknown language; the letters are divided into words—all wrongly, of course—these words are analysed with a great show of erudition, and a literal Latin version accompanies the whole. If I remember (for I have mislaid the amusing production) it proves to be an invocation of the gods, to be used at a sacrifice. Now, a joke is a good thing in its place; only do not let it be made too much of. Every archæologist, beginning with Jonathan Oldbuck, must sometimes fall into blunders, when he takes inscriptions in hand, even if the language be a known one; and, of course, *à fortiori*, when

---

[1] *Monumental History of Egypt*, Vol. I. p. 262. London, 1854.

but little known. My own opinion on hieroglyphics would be of no value whatever, as I know nothing beyond what I have read in a few modern authors, and have never studied the subject; but, allow me to observe, that I had a conversation very lately with my learned and excellent friend, Dr Birch, of the British Museum, who is now engaged in making a dictionary of hieroglyphics, and he assured me that a real progress has been made in the study of them, that a great deal of certainty has been attained to; while there is still much that requires further elucidation. To the judgment of such a man, who has spent a great part of his life in the study of Egyptian antiquities, though he has splendidly illustrated other antiquities also, I must think that greater weight should be attached than to the judgment of others, eminent as they may be in some branches of learning, who have never studied this as a specialty.

The relation of archæology to Egyptian history deserves especial notice. We have not here, as in pre-historic Europe, a mere multitude of uninscribed and inconsiderable remains; but we have colossal monuments of all kinds—temples, gateways, obelisks, statues, rock sculptures—more or less over-written with hieroglyphics; also sepulchral-chambers, in many instances covered with paintings, in addition to a variety of smaller works, mummy cases, jewelry, scarabæi, pottery, &c., upon many of which are inscriptions. By aid of these monuments mostly, but by no means exclusively, the history of the Pharaohs and the manners and customs of their people are recovered. The *monumenta litterarum* themselves are frequently preserved on the monuments of stone and other materials.

For the pyramids of Ghizeh and the adjoining districts, for the glorious temples of Dendera, of Karnak, the grandest of all the remains of the Pharaohs, as well as for those of Luxor, with its now one obelisk, of Thebes, of Edfou, of Philæ, likewise for the grottoes of Benihassan, I must leave you to your own imagination or recollection, which may be aided in some degree by a few of the beautiful photographs by Bedford, which are now before your eyes. They extend along the banks and region of the Nile—for this is Egypt—from the earliest times down to the age of the Ptolemies and of Cleopatra herself, and even of the Roman empire, in the case of Dendera, where the portico was added by Tiberius to Cleopatra's temple. Before quitting these regions I would remark, that the extraordinary rock-hewn temple of Aboo-Simbel in Nubia, which includes the most beautiful colossal statues yet found—their height as they sit is more than fifty feet—bears some similarity to certain Indian temples, especially to the temple of Siva at Tinnevelly, and the Kylas at Ellora, which last has excited the astonishment of all travellers. "Undoubtedly," says Mr Fergusson, "there are many very striking points of resemblance...but, on the other hand, the two styles differ so widely in details and in purpose, that we cannot positively assert the actual connexion between them, which at first sight seems unquestionable[1]."

The archæology of the Babylonian empire need only occupy a few moments. The antiquity of Babylon is proved to be as remote as the fifteenth century B.C., by the occurrence of the name on a monument of Thothmes III., an Egyptian monarch of the XVIIIth dynasty. It may be much older than that; but the archæological remains of the Birs Nimroud (which was long imagined to be the tower of Babel) hitherto found are not older than the age of Nebuchadnezzar. This palatial structure consisted, in Mr Layard's opinion, of successive horizontal terraces, rising one above another like steps in a staircase. Every inscribed brick taken from it,—and there are thousands and tens of thousands of these,—bears the name of Nebuchadnezzar. It is indeed possible that he may have added to an older structure, or rebuilt it; and if so we may one day find more ancient relics in the Birs. But at a place called Mujelibé (the Overturned) are remains of a Babylonian palace not covered by soil, also abounding with Nebuchadnezzar's bricks, where Mr Layard found one solitary fragment of a sculptured slab, having representations of gods in head-dresses of the Assyrian fashion, and indicating that the Baby-

[1] *Handbook of Architecture*, p. 101. London, 1859.

lonian palaces were probably similarly ornamented. A very curious tablet was also brought from Bagdad of the age of Nebuchadnezzar, giving, according to Dr Hincks, an account of the temples which he built. Besides these, "a few inscribed tablets of stone and baked clay, figures in bronze and terra cotta, metal objects of various kinds, and many engraved cylinders and gems are almost the only undoubted Babylonian antiquities hitherto brought to Europe." Babylonia abounds in remains, but they are so mixed—Babylonian, Greek, Roman, Arsacian, Sassanian, and Christian —that it is hard to separate them. Scarcely more than one or two stone figures or slabs have been dug out of the vast mass of débris; and, as Isaiah has said, "Babylon is fallen, is fallen; and all the graven images of her gods hath Jehovah broken unto the ground[1]."

The most splendid archæological discovery of our age is the disinterment of the various palaces and other monuments of the Assyrian Empire. The labours of Mr Layard and M. Botta have made ancient Assyria rise before our eyes in all its grandeur and in all its atrocity. In visiting the British Museum we seem to live again in ancient Nineveh. We behold the sculptured slabs of its palaces, on which the history of the nation is both represented and written; we wonder at its

[1] See Layard's *Nineveh and Babylon*, chapters xxii. xxiii., especially pp. 504, 528, 532. London, 1853.

strange compound divinities, its obelisks, its elegant productions in metal, in ivory, and in terra cotta. By patient and laborious attention to the cuneiform inscriptions, aided by the notices in ancient authors, sacred and profane, men like Sir H. Rawlinson and Dr Hincks have recovered something like a succession of Assyrian kings, ranging from about 1250 B.C. to about 600 B.C., and many particulars of their reigns, some of which bring out in a distinct manner the accurate knowledge of the writers of the Old Testament.

The remains of ancient Persia are too considerable to be passed over. Among other monuments at Pasargadæ, a city of the early Persians, is a great monolith, on which is a bas-relief, and a cuneiform inscription above, "I am Cyrus the king, the Achæmenian." Here is the tomb of the founder of the empire.

At Susa, the winter seat of the Persian kings from the time of Cyrus, Mr Loftus and Sir W. F. Williams have found noble marble structures raised by Darius, the son of Hystaspes (424—485 B.C.), whose great palace was here: commenced by himself and completed by Artaxerxes II. or Mnemon (405—359 B.C.). Both here and at Persepolis, the richest city after Susa (destroyed, as we all remember from Dryden's ode, by Alexander), are ruins of magnificent columns of the most elaborate ornamentation, and many cuneiform inscriptions, deciphered by Lassen and Rawlinson. Mr Loftus remarks on the great similarity of the buildings of Persepolis and Susa, which form a distinct style of architecture. This is the salient feature of Persian archæology, and to him I refer you upon it[1]. I cannot dwell upon other ruins in these regions, or on the minor objects, coins, cylinders, and vases of the ancient Persian empire; and still less on the very numerous coins of the Arsacidæ, and Sassanidæ, who afterwards succeeded to it.

Of ancient Judæa we possess as yet very scanty archæological monuments indeed before the fall of the monarchy. The so-called Tombs of the Kings are now, I believe, generally considered to belong to the Herodian period. Of the Temple of Jerusalem, the holy place of the Tabernacle of the Most Highest, not one stone is left upon another. And we may well conceive that nothing less than its destruction would effectually convince the world of the great truth that an hour had arrived in which neither that holy mountain on which it was built, nor any other in the whole world, was to be the scene of the exclusive worship of the Father. The sites of the Holy Places, however, have naturally excited much attention, and have been well illustrated by several distinguished resident members of our University, and also by a foreign gentleman who for some time resided among us. Dr Pierotti had the singular

---

[1] See his *Travels and Researches in Chaldæa and Susiana*, ch. xxviii. London, 1857; also Smith's *Dict. of Greek and Roman Geography*, s. v. Pasargadæ, Persepolis, Susa; and Vaux's *Nineveh and Persepolis*. London, 1850.

good fortune to discover the subterranean drains by which the blood of the victims, slaughtered in the Temple, was carried off; and this discovery afforded valuable aid in determining various previously disputed matters in connexion with the Temple. He likewise came upon some masonry in the form of bevelled stones below the surface, which was not unreasonably supposed to belong to Solomon's Temple; but it now appears that this opinion is doubtful. Besides these, we have the sepulchres of the patriarchs at Hebron, guarded with scrupulous jealousy; and tanks at the same place, which may be as old as the time of David, and perhaps one or two things more of a similar kind. We may well hope that the explorations which are now being set on foot for bringing to light the antiquities of Palestine may add to their number.

In the relation of Jewish archæology to Jewish history we have a case quite different to all those that have gone before it: there the native archæology was more or less extensive, the independent native literature scanty or non-existent; here, where the archæology is almost blotted out, is it precisely the reverse. We have in the sacred books of the Old Testament an ample literary history: we have scarcely any monumental remains of regal Judæa at all. With regard to the New Testament the matter is otherwise; archæological illustrations, as well as literary, exist in abundance, and some very striking proofs from archæology have been adduced of the veracity and trustworthiness of its authors. My predecessor bestowed great attention on the numismatic and other monumental illustrations of Scripture, and herein set a good example to all that should come after him. Archæology is worthily employed in illustrating every kind of ancient literature; most worthily of all does she occupy herself in the illustration and explanation and confirmation of the sacred writings, of the Book of books.

The antiquities of Phœnicia need not detain us long. Opposite to Aradus is an open quadrangular enclosure, excavated in rock, with a throne in the centre for the worship of Astarte and Melkarth; this is the only Phœnician temple discovered in Phœnicia, except a small monolithal temple at Ornithopolis, about nine miles from Tyre, of high antiquity, dedicated apparently to Astarte. I wish however to direct your attention to the characteristic feature of Phœnician architecture, its enormous blocks of stone bevelled at the joints. You have them in the walls of Aradus and in other places in Phœnicia. They are also found in the temple of the Sun at Baalbec, and may with great probability, I conceive, be regarded as Phœnician; though the rest of the beautiful architectural remains there are Greco-Roman of the Imperial period, and perhaps the best specimens of their kind in existence. Among other Phœnician antiquities we have sarcophagi, and sepulchral chambers for receiving them, also very beautiful variegated glass found over a good part of Europe and Asia, commonly called Greek, but perhaps more reasonably presumed to be Phœnician. Most of the remains found on the sites of the Phœnician settlements are either so late Phœnician, or so little Phœnician at all, as at Carthage, that I shall make no apology for passing over both them, and the few exceptions also, just alluding however to the existence of a remarkable hypæthral temple in Malta, which I myself saw nearly twenty years ago, not long, I believe, after it was uncovered. With regard to the strange vaulted towers of Sardinia, called Nuraggis, they may be Phœnician or Carthaginian, but their origin is uncertain. "All Phœnician monuments," says Mr Kenrick, "in countries unquestionably occupied by the Phœnicians are recent[1]." He makes the remark in reference to the Lycian archæology. Whether the Lycians were of Phœnician origin or not, their rock-temples and rock-tombs, abounding in sculptures (illustrative both of their mythology and military history), shew that they were not much behind the Greeks in the arts. With the general appearance of their Gothic-like architecture, and of their strange bilingual inscriptions, Greek and Lycian, we are of course familiarised by the Lycian Room in the British Museum. With regard to the relation of Phœnician and Lycian archæology to the history

[1] *Phœnicia*, p. 88. London, 1855. See also Smith's *Dict. of Greek and Roman Geography*, s. v. Phœnicia and Lycia.

of the peoples themselves, it must be sufficient to say, that their history, both literary and monumental, is quite fragmentary; in the case of Phœnicia the literary notices perhaps preserve more to us than the monumental; in regard to Lycia the remark must rather be reversed.

From Phœnicia, which first carried letters to Greece, let us also pass to Greece. But Greece, in the sense in which I shall use it, includes not only Greece Proper, but many parts of Asia Minor, as well as Sicily and the Great Greece of Italy. And here I must unwillingly be brief, and make the splendid extract from Canon Marsden, quoted before, in some degree do duty for me. But think for a minute first on its architecture, I do not mean its earliest remains, such as the Cyclopian walls and the lion-gate at Mycenæ, and the so-called treasury of Atreus, which ascend to the heroic ages or farther back, but its temple architecture. Before I can name them, images of the Parthenon, the Erectheum, the temple of Jupiter Panhellenius at Ægina, the temple of Apollo Epicurius at Phigalia or Bassæ, that of Concord (so called) at Agrigentum, the most perfect in Sicily, the three glorious Doric temples of Pæstum, the Ionic ruins of Branchidæ, will, I am confident, have arisen before your eyes. Many of us perhaps have seen some of them; if not, we all feel as though we had. Think of its sepulchral monuments, which are in the form of temples; and first of Queen Artemisia's Mausoleum, the most splendid architectural expression of conjugal affection that has ever existed, the wonder of the world, with its colossal statue of her husband and its bas-reliefs by Bryaxis and Scopas and other principal sculptors; and remember that we have these in our national museum. Various fine rock-tombs, likewise in the form of temples, occur in Asia Minor, *e.g.* that of Midas at Nacoleia, the Lion-tomb at Cnidus, the necropolis at Telmessus.

The transition from temples and tombs to statuary is easy, as these were more or less decorated with its aid. Although we still possess the great compositions of some of the first sculptors and brass-casters, for example, the Quoit-thrower of Myron, the Diadumenos of Polycleitus, (*i.e.* a youth binding his head with a fillet in token of an athletic victory,) and perhaps several of the Venuses of Praxiteles; yet it is needless for me to remind you that these with few exceptions are considered to be copies, not originals. But yet there are exceptions. "The extant relics of Greek sculpture," says Mr Bunbury, "few and fragmentary as they undoubtedly are, are yet in some degree sufficient to enable us to judge of the works of the ancient masters in this branch of art. The metopes of Selinus, the Æginetan, the Elgin, and the Phigaleian marbles, to which we now add the noble fragments recently brought to this country from Halicarnassus, not only serve to give us a clear and definite idea of the progress of the art of sculpture, but enable us to estimate for ourselves the mighty works which were so celebrated in antiquity[1]." Of bronzes of the genuine Greek period, which we may call their metal statuary, the most beautiful that occur to my remembrance are those of Siris, now in the British Museum. They are considered by Brönsted to agree in the most remarkable and striking manner with the distinctive character of the school of Lysippus. But most of the extant bronzes are, I believe, of the Roman period, executed however, like their other best works, by Greco-Roman artists.

With the Greek schools of painting, Attic, Asiatic, and Sicyonian, no less celebrated than their sculpture, it has fared far worse. There is not one of their works surviving; no, not one. Of these schools and their paintings I need not here say anything, as I am concerned only with the archæological monuments which are now in existence. But the loss is compensated in some degree by the paintings on vases, in which we may one day recognise the compositions of the various great masters of the different schools, just as in the majolica and other wares of the

---

[1] *Edinburgh Review* for 1858, Vol. CVIII. p. 382. I follow common fame in assigning this article to Mr Bunbury; few others indeed were capable of writing it. Besides the sculptures named by him we have in the British Museum a bas-relief by Scopas, as it is thought, who may also be the author of the Niobid group at Florence; likewise the Ceres (so-called) from Eleusis, and the statue of Pan from Athens, now in our Fitzwilliam Museum. For other antique statues and bronzes and for the later copies see Müller's *Ancient Art*, passim.

16th and following centuries we have the compositions of Raffaelle, Giulio Romano, and other painters. "The glorious art of the Greek painters," says K. O. Müller, the greatest authority for ancient art generally, "as far as regards light, tone, and local colours, is wholly lost to us; and we know nothing of it except from obscure notices and later imitations;" (referring, I suppose, to the frescoes of Herculaneum and of Pompeii more especially;) "on the contrary, the pictures on vases with thinly scattered bright figures give us the most exalted idea of the progress and achievements of the art of design, if we venture, from the workmanship of common handicraftsmen, to draw conclusions as to the works of the first artists[1]." But of this matter and of the vases themselves, which rank among the most graceful remains of Greek antiquity, and are found over the whole Greek world, I shall say no more now, as they will form the subject of my following lectures. We have also many terra cottas of delicate Greek workmanship, mostly plain, but some gilded, others painted, from Athens, as well as from a great variety of other places, of which the finest are now at Munich. Relief ornaments, sometimes of great beauty, in the same material, were impressed with moulds, and Cicero, in a letter to

[1] *Ancient Art and its Remains,* p. 119. Translated (with additions from Welcker) by Leitch. London, 1852. This invaluable work is a perfect thesaurus for the student, and will conduct him to the most trustworthy authorities on every branch of the subject.

Atticus, wishes for such *typi* from Athens, in order to fix them on the plaster of an atrium. Most of those which now remain seem to be of Greco-Roman times.

Of the art of coinage invented by the Greeks and carried by them to the highest perfection which it has ever attained, a few words must now be said. The history of a nation, said the first Napoleon, is its coinage: and the art which the Greeks invented became soon afterwards, and now is, the history of the world. Numismatics are the epitome of all archæological knowledge, and any one who is versed in this study must by necessity be more or less acquainted with many others also. Architecture, sculpture, iconography, topography, palæography, the public and private life of the ancients and their mythology, are all illustrated by numismatics, and reciprocally illustrate them.

Numismatics give us also the succession of kings and tyrants over the whole Greek world. In the case of Bactria or Bactriana, whose capital Bactra is the modern Balk, this value of numismatics is perhaps most conspicuous. From coins, and from coins almost alone, we obtain the succession of kings, beginning with the Greek series in the third century B.C., and going on with various dynasties of Indian language and religion, till we come down to the Mohammedan conquest. "Extending through a period of more than fifteen centuries," says Professor H. H. Wilson, "they furnish a distinct outline of the great political and religious vicissitudes of an important division of India, respecting which written records are imperfect or deficient[1]."

Coins are so much more durable than most other monuments, that they frequently survive, when the rest have perished. This is well put by Pope in his Epistle to Addison, on his Discourse on Medals:

> Ambition sighed, she saw it vain to trust
> The faithless column and the crumbling bust,
> Huge moles whose shadows stretched from shore to shore,
> Their ruins perished and their place no more.
> Convinced she now contracts her vast design,
> And all her triumphs shrink into a coin.
> A narrow orb each crowded conquest keeps,
> Beneath her palm here sad Judæa weeps;
> Now scantier limits the proud arch confine;
> And scarce are seen the prostrate Nile or Rhine;
> A small Euphrates thro' the piece is rolled,
> And little eagles wave their wings in gold.
> The Medal, faithful to its charge of fame,
> Through climes and ages bears each form and name;
> In one short view subjected to our eye,
> Gods, emperors, heroes, sages, beauties, lie.

Regarded simply as works of art the coins of Magna Græcia and Sicily, more especially those of Syracuse and its tyrants, as well as those of Thasos, Opus, and Elis, also the regal coins of Philip, Alexander, Mithridates, and some of the Seleucidæ, are amongst the most exquisite produc-

---
[1] *Ariana Antiqua*, p. 439. London, 1841. For the more recent views of English and German numismatists on these coins, see Mr Thomas's *Catalogue of Bactrian Coins* in the Numismatic Chronicle for 1857, Vol. XIX. p. 13 sqq.

tions of antiquity. Not even in gem-engraving, an art derived by Greece from Egypt and Assyria, but carried by her to the highest conceivable perfection, do we find anything superior to these. I must, before quitting the subject of numismatics, congratulate the University on the acquisition of one of the largest and most carefully selected private collections of Greek coins ever formed, viz. the cabinet of the late Col. Leake, which is now one of the principal treasures of the Fitzwilliam Museum.

Inferior as gems are to coins in most archæological respects, especially in respect of their connection with literary history, and though not superior to the best of them artistically, gems have nevertheless one advantage over coins, that they are commonly quite uninjured by time. Occasionally (it is true) this is the case with coins; but with gems it is the rule. Of course, to speak generally, the art of gems, whose material is always more or less precious, is superior to that of coins, which were often carelessly executed, as being merely designed for a medium of commercial exchange. High art would not usually spend itself upon small copper money, but be reserved for the more valuable pieces, especially those of gold and silver[1]. The subjects of gems are mostly mythological, or are connected with the heroic cycle; a smaller, but more inter-

[1] This remark however must not be pressed too closely. Certain small Greek copper coins of Italy, Sicily, &c., are exceedingly beautiful.

esting number, presents us with portraits, which however are in general uninscribed. At the same time, by comparing these with portrait-statues and coins we are able to identify Socrates, Plato, Aristotle, Demosthenes, Alexander the Great, several of the Ptolemies, and a few others; most of which may have been engraved by Greco-Roman artists. But the catalogue of authentic portraits preserved to us, both Greek and Roman, is, as K. O. Müller observes, now very much to be thinned.

With regard to ancient iconography in general, coins, without doubt, afford the greatest aid; but no certain coin-portraits are, I believe, earlier than Alexander[1]. The oldest Greek portrait-statue known to me is that of Mausolus, now in the British Museum; but the majority of the statues of Greek philosophers and others are probably to be referred to the Roman times, when the formation of portrait-galleries became a favourite pursuit. With the Greeks it was otherwise; the ideal was ever uppermost in their mind: they executed busts of Homer indeed and placed his head on many of their coins; but of course these were no more portraits than the statues of Jupiter and Pallas are portraits. With regard to the relation of Greek archæology to the history of Greece, both the monuments and the literature are

[1] I am aware that there are reasons for believing that a Persian coin preserves a portrait of Artaxerxes Mnemon, who reigned a little earlier.

abundant, and they mutually illustrate one another; and the same remark is more or less true for the histories of the nations afterwards to be mentioned, upon which I shall therefore not comment in this respect.

From Greece, who taught Rome most or all that she ever knew of the arts, we pass to the contemplation of the mistress of the world herself. She found indeed in her own vicinity an earlier civilisation, the Etruscan, whose archæological remains and history generally are amongst the most obscure and perplexing matters in all the world of fore-time. The sepulchral and other monuments of Etruria are often inscribed, but no ingenuity has yet interpreted them. The words of he Etruscan and other Italian languages have been recently collected by Fabretti. There is some story about a learned antiquary after many years' research coming to the conclusion that two Etruscan words were equivalent to *vixit annos*, but which was *vixit*, and which *annos*, he was as yet uncertain. We have also Etruscan wall-paintings, and various miscellaneous antiquities in bronze, and among them the most salient peculiarity of Etruscan archæology not easily to be conjectured, its elegantly-formed bronze mirrors. These, which are incised with mythological subjects, and often inscribed, have attracted the especial attention of modern scholars and antiquaries, who have gazed upon them indeed almost as wistfully as the Tuscan ladies themselves.

But Greece had far more influence over Roman life and art than Etruria.

> Græcia capta ferum victorem cepit, et artes
> Intulit agresti Latio.

Accordingly, Greek architecture (mostly of the later Corinthian style, which was badly elaborated into the Composite) was imported into Rome itself, and continued to flourish in the Greek provinces of the empire. Temples and theatres continued much as before; but the triumphal arch and column, the amphitheatre, the bath and the basilica, are peculiarly Roman.

The genius of Rome however was essentially military, and the stamp which she has left on the world is military also. Her camps, her walls, and her roads, *strata viarum*, which, like arteries, connected her towns one with another and with the capital, are the real peculiarities of her archæology. The treatise on Roman roads, by Bergier, occupies above 800 pages in the *Thesaurus* of Grævius. Instead of bootlessly wandering over the width of the world on these, let us rather walk a little over those in our own country, and as we travel survey the general character of the Roman British remains, which may serve as a type of all. In the early part of this lecture, I observed that we, in common with the rest of Western Europe, find in our islands weapons which belong to the stone, bronze, and iron periods; and here also, as in other places, the last-named period doubtless connects itself with the Roman. But besides these, we have other remains, many of which may be referred to the Celtic population which Cæsar had to encounter, when he invaded our shores. These remains may in great part perhaps (for I am compelled to speak hesitatingly on a subject which I have studied but little, and of which no one, however learned, knows very much) be anterior to Roman times. Of this kind are the cromlechs at Dufferin in South Wales, in Anglesey, and in Penzance, of which there are models in the British Museum; of this kind also are, most probably, the gigantic structures at Stonehenge, about which so much has been written and disputed. The British barrows of various forms and other sepulchral remains may also be referred, I should conceive, in part at least, to the pre-Roman Celtic period. The earlier mounds contain weapons and ornaments of stone, bronze and ivory, and rude pottery; the later ones, called Roman British barrows, appear mostly not to contain stone implements, but various articles of bronze and iron and pottery; also gold ornaments and amber and bead necklaces. Other sepulchral monuments consist merely of heaps of stones covering the body which has been laid in the earth. Many researches into this class of remains have of late years been made, and by none perhaps more patiently and more successfully than by the late Mr Bateman, in Derbyshire. The archæology of Wales has also been made the special object of study by a society formed for the purpose.

Some tribes of the ancient Britons were certainly acquainted with the art of die-sinking, and a great many coins, principally gold, are extant, some of which may probably be as early as the second century before Christ. They are, to speak generally, barbarous copies of the beautiful gold staters of Philip of Macedon, which circulated over the Greek world, and so might become known to our forefathers by the route of Marseilles.

With these remarks I leave the Celtic remains in Britain; all attempts to connect together the literary notices and the antiquities of the Celts and Druids, so as to make out a history from them, have been compared to attempts to "trace pictures in the clouds¹." Still we may say to the Celtic archæologist,

Οαρσείν χρή, φίλε Βάττε, τάχ' αύριον ἔσσετ' ἄμεινον.

One day matters may become clearer by the help of an extended and scientific archæology.

But of the Romano-British remains it may be necessary to say something. When we look at the map in Petrie's *Monumenta Historica Britannica*, in which the Roman roads are laid down by their actual remains, we see the principal Roman towns and stations connected together by straight lines, which are but little broken. So numerous are they that we might almost fancy that we were looking at a map in an early edition of a

¹ *Pict. Hist. of England*, Vol. I. p. 59. London, 1837.

Railway Guide. In this county they abound and have been very carefully traced, and both here and in other counties are still used as actual roads. In a few instances mile-stones have also been found. In our own country, cut off, as Virgil says, from the whole world, we do not expect the splendid monuments of Roman greatness, yet even here the temple, the amphitheatre and the bath are not unknown; and in our little Pompeii at Wroxeter we have, if my memory deceive me not, some vestiges of fresco-painting, an art of which we have such beautiful Roman examples elsewhere. But everywhere we stumble upon camps and villas; everywhere

> The tesselated pavements shew
> Where Roman lamps were wont to glow.

And of these lamps themselves we have an infinite number and variety, and on many of them representations of the games of the circus and of various other things, formed in relief; a remark which may also be made of their fine and valuable red Samian ware; fragments of which are commonly met with, but the vases are rarely entire. Of their other pottery, and of their glass and personal ornaments, and miscellaneous objects, I must hardly say any thing; but only observe that the Romans have left us a very interesting series of coins relating to Britain; Claudius records in gold the arch he raised in triumphant victory over us: in the same way Hadrian, Antoninus Pius, Septimius Severus, besides building their great walls against us, have, as well as Caracalla and Geta, struck many pieces in silver and copper to commemorate our tardy subjugation. The British emperors or usurpers, Carausius and Allectus, have also left us very ample series of coins, and indeed it is by these, much more than by the monuments of letters, that their histories are known. In the fourth and fifth centuries the monetary art declined greatly in the Western Empire, and was on the whole at a very low ebb in the Eastern or Byzantine Empire, and in the middle ages, generally, throughout Europe.

At Constantinople a new school of Roman art arose, which exercised a powerful influence on medieval art in general. Soon after the foundation of Constantinople, Roman artists worked there in several departments with a skill by no means contemptible, though of a strangely conventional and grotesque character; and from them, as it would seem, the medieval artists of Central and Western Europe caught the love of the same crafts, and carried them to much higher excellence. I would allude in the first place, as being among the earliest, to ivory carvings, principally consular diptychs. From the time of the emperors it was the custom for consuls and other curule magistrates to make presents both to officials and their friends of ivory diptychs, which folded together like a pair of book-covers, on which sculptures in low relief were carved, as a mode of announcing their elevation. From the fourth and fifth centuries down to the fourteenth we find them, some of the earliest with classical subjects, as the triumph of Bacchus, probably of the fourth century; but mostly with Scriptural ones, or with representations of consuls. Some of these are enriched with jewellery. The inscriptions accompanying them are either in Greek or in Latin. In Germany they occur in the Carlovingian period, though rarely, and in France and Italy later still. Perhaps it should be mentioned that the ivory episcopal chair of St Maximian at Ravenna, a work of the sixth century, is the finest example extant of this class of antiques, and is doubly interesting as being one of the very few extant specimens of furniture during the first three centuries of the middle ages. Various casts of medieval ivories, it may be added, have been executed and circulated by the Arundel Society.

Another art learnt from Rome in her decline, or from Constantinople, is the illumination of MSS., which the calligraphers of the middle ages in all countries throughout Europe carried to a very high perfection. Perhaps the earliest example to be named is the Greek MS. of Genesis in the LXX, now preserved in the Imperial Library at Vienna, probably of the fourth century. The vellum is stained purple, and the MS. is decorated with pictures executed in a quaint, but vigorous style. In these, we find (as M. Labarte[1], a great

[1] *Histoire des Arts au moyen âge.* Album. Vol. II. pl. lxxvii. Paris, 1864.

authority for medieval art, assures us) all the characters of Roman art in its decline, such as it was imported to Constantinople by the artists whom Constantine called to his new capital; and "they have served," as he adds, "for a point of departure" in the examination which he has made of the tendencies and destinies of Byzantine art. Compare the Vatican MSS. of Terence and Virgil. I cannot be expected to enter into details about illuminations; they occur in MSS. of all sorts, more or less, in Europe, down to the sixteenth century, but especially in sacred books, such as were used in Divine service. I need only call to your remembrance the beautiful assemblage exhibited in the Fitzwilliam Museum and in the University Library, to say nothing of the treasures possessed by our different colleges.

There are many other objects of medieval art not unworthy of being enlarged upon, which I intentionally pass over lightly, lest their multiplicity should distract us; thus I will say little of its pottery, its coins, or of its sculptures and bas-reliefs in stone. With regard to the first of them, M. Labarte observes: "It is not until the beginning of the fifteenth century that we find among the European nations any pottery, but such as has been designed for the commonest domestic use, and none that art has been pleased to decorate." These are objects which the middle ages have in common with others; and they are objects in which a comparison will not be favourable to medieval art. Still, we must take care that a love of art does not blind us to the real value of such things; they are always interesting for the *history* of art, whatever their rudeness or whatever their ugliness; and, moreover, they are often, as the coins of various nations, of high historical interest. For example, on our own series of barbarous Saxon coins we have not only the successions of kings handed down to us, in the several kingdoms of the so-called Heptarchy and in the united kingdom, but also on the reverses of the same coins we have mention made of a very large number of cities and towns at which they were respectively struck. For example, to take Cambridge, we find that coins were struck here by King Edward the Martyr, Ethelred the Second, Canute, Harold the First, and Edward the Confessor; also after the Conquest by William the First and William the Second. We are thus furnished with very early notices, and so in some measure able to estimate the importance of the cities and towns of our island in medieval times; though great caution is necessary here in making deductions; for no coins appear to have been struck in Cambridge after the reign of William Rufus. And this seems at first sight so much the more surprising when we bear in mind that money was struck in some of our cities, as York, Durham, Canterbury, and Bristol, quite commonly, as late as the fifteenth and sixteenth centuries. But, in truth, from the twelfth century downwards, the number of cities and towns in which lawful money was struck became comparatively small.

But I must not wander too far into numismatics. The art of enamelling, peculiarly characteristic of the later periods of the middle ages, is very fully treated of by M. Labarte, from whom I derive the following facts. The most ancient writer that mentions it is the elder Philostratus, a Greek writer of the third century, who emigrated from Athens to Rome. In his *Icones*, or *Treatise on Images*, the following passage occurs. After speaking of harness enriched with gold, precious stones, and various colours, he adds: "It is said that the barbarians living near the ocean pour colours upon heated brass, so that these adhere and become like stone, and preserve the design represented." It may, therefore, be considered as established that the art of enamelling upon metals had no existence in either Greece or Italy at the beginning of the third century; and, moreover, that this art was practised at least as early in the cities of Western Gaul. During the invasions and wars which desolated Europe from the fourth to the eleventh century almost all the arts languished, and some may have been entirely lost. Enamelling was all but lost; for between the third and the eleventh centuries the only two works which occur as landmarks are the ring of King Ethelwulf in the British Museum, and the ring of Albstan, probably the bishop of Sherburne, who lived at the same time. These two little pieces, however, only serve to establish the bare existence of enamelling in the West in the ninth century. But in this same century the art was in all its splendour at Constantinople, and we possess specimens of Byzantine workmanship of even an earlier date. I cannot enter into the various modes of enamelling, which are fully described by M. Labarte; but merely mention, without comment, a few of the principal specimens, independently of the Limoges manufacture, which constituted the chief glory of that city from the eleventh century to the end of the medieval period. "This became the focus whence emanated nearly all the beautiful specimens of enamelled copper, which are so much admired and so eagerly sought after for museums and collections." The principal earlier examples then are these; the crown and the sword of Charlemagne, of the ninth century, now in the Imperial Treasury at Vienna; the chalice of St Remigius, of the twelfth century, in the Imperial Library at Paris; the shrine of the Magi in Cologne, and the great shrine of Nôtre Dame at Aix-la-Chapelle, presented by the Emperor Frederick Barbarossa in the latter part of the same twelfth century. Also the full-length portrait (25 inches by 13) of Geoffrey Plantagenet, father of our Henry II., which formerly ornamented his tomb in the cathedral, but is now in the Museum at Le Mans. The British Museum likewise contains two or three fine examples; and among them an enamelled plate representing Henry of Blois,

Bishop of Winchester, and brother of King Stephen.

Very fine also are the extant products of the goldsmith's art in the middle ages; which date principally from the eleventh century, when the art received a new impulse in the West; those of earlier date, with very few exceptions, now cease to exist. They are principally chalices, reliquaries, censers, candlesticks, croziers and statuettes.

Nor can I pass over in absolute silence the armour of the middle ages. Until the middle of the ninth century it would appear to have resembled the Roman fashion, of which it is needless to say anything; but in Carlovingian times the hilts and scabbards of dress-swords were very highly decorated; and about this period, or rather later, the description of armour used by the ancients was exchanged for the hauberk or coat of mail, which was the most usual defensive armour during the period of the Crusades. The first authentic monument where this mail-armour is represented is on the Bayeux tapestry of Queen Matilda, representing the invasion of England by William Duke of Normandy in 1066; the most famous example of medieval tapestry in existence, though other specimens are to be seen at Berne, Nancy, La Chaise Dieu, and Coventry. The art of the *tapissier*, however, in the eleventh century, when the Bayeux tapestry was made, would appear to have been on the decline.

In the beginning of the fourteenth century plate-armour began to come into use; and by and by this was decorated with Damascene work, a style of art applied to the gate of a basilica in Rome, which was sent from Constantinople, as early as the eleventh century, but which did not become general in the West till the fifteenth. To this I may just add, that sepulchral brasses, on which figures in armour are often elaborately represented by incised lines, are a purely medieval invention of the thirteenth century. Sir Roger de Trumpington's brass at Trumpington is one of the very earliest examples. But time forbids me to say more of sepulchral brasses, a class of antiquities almost confined to our own country, of which we have some few specimens as late as the seventeenth century, or to do more than allude to the beautiful sepulchral monuments in stone of the medieval period, with which we are all more or less familiar.

The most remarkable art to which the middle age gave birth was oil-painting, the very queen of all the fine arts, though it was to the age of the Medici that its immense development was due. Previously painting had been subordinated to architecture; but now, while mosaics, frescoes, and painted glass remained still subservient to her, the art of painting occupies a distinct and prominent rank of its own. It used commonly to be said that the invention of painting on prepared panel was due to Margaritone of Arezzo, who died about 1290, and in like manner that John van Eyck invented oil-painting in 1410. Both these errors have been propagated by the authority of Vasari. But it is now well known, and has been conclusively proved, both by M. Labarte and by Sir C. Eastlake, that these modes of painting are mentioned by authors who lived more than a century before Margaritone, in particular by the monk Theophilus, who in the twelfth century composed a work entitled *Diversarum artium schedula*. Paintings in oil either are or lately were in existence anterior to John van Eyck; for example one at Naples, executed by Filippo Tesauro, and dated 1309. We must ascend to much earlier times to discover the true origin of portable paintings, and we shall find it in the Byzantine Empire. The Greeks, about the time that the controversy respecting images was rife, multiplied little pictures of saints; these were afterwards brought over in abundance by the priests and monks who followed the crusades, and from the study of them, schools of painting in tempera arose in Italy, in the twelfth century, at Pisa, Florence and other places. The Byzantine school, M. Labarte tells us, reigned paramount in Italy until the time of Giotto, *i. e.* the beginning of the fourteenth century, and also in the schools of Bohemia and Cologne, the most ancient in northern Europe, until towards the end of the fourteenth century. In this country we have two very early paintings, one of the beginning and the other of the end of the same fourteenth century, in Westminster Abbey. The former, probably a decoration of the high altar, is on wood; it represents the Adoration of the Magi and other Scriptural subjects, and is declared by Sir C. Eastlake to be worthy of a good Italian artist of the fourteenth century, though he thinks that it was executed in England. The latter is the canopy of the tomb of Richard II. and Anne his first wife, representing the Saviour and the Virgin and other figures. The action and expression are declared by Sir C. Eastlake to indicate the hand of a skilful painter. In 1396, £20 was paid by the sacrist for the execution of the work. These remarks must suffice for a notice of medieval painting; the glorious period of its history belongs rather to the Renaissance, or post-medieval age.

The only archæological monuments of great importance which remain to be mentioned are those of architecture, in connection with the accessories of mosaics, frescoes, and painted glass. The two former descended from classical times, the last is the creation of the middle age. Mosaics having been originally used only in pavements, at length were employed as embellishments for the walls of basilicas, and, by a natural transition, of churches. Constantine and his successors decorated many churches in this manner, and in the East a ground of gold or silver was introduced below the glass cubes of the mosaics, and a lustre was by this means spread over the work which in earlier times was altogether unknown. Thus the tympanum above the principal door of the narthex of the Church of St Sophia, built by the Emperor Justinian at Constantinople, is adorned with a mosaic picture of the Saviour seated, the cubes of the mosaics being of silvered glass; it is accompanied by Greek texts. This and other later mosaics are figured by M. Labarte, in his last and most splendid work, entitled *Histoire des Arts au moyen âge;* among the rest a Transfiguration of the tenth century. The Byzantine art, with its stiff conventionality, prevailed every where till Cimabue, G. Gaddi, and Giotto imparted to its rudeness a grace and nobleness which marked a new era. In the vestibule of St Peter is a noble mosaic, partly after the design of Giotto, representing Christ walking on the water, and the apostles in the ship. But the very masters who raised the art to its perfection brought about its destruction. Painting, restored by these same great men, was too powerful a rival; and after the sixteenth century, when it still flourished in Venice under the encouragement of Titian, we hear little more of mosaics on any great scale.

Passing over frescoes, which were much encouraged by Charlemagne, and by various sovereigns and popes during the middle ages, because the ravages of time have either destroyed them altogether or left them in a deplorable condition, as for example in some parish-churches in England, I will make a few remarks on painted glass, so extensively used in the decoration of the later churches.

The art of painting glass was unknown to the ancients, and also to the early periods of the middle ages. "It is a fact," says M. Labarte, "acknowledged by all archæologists, that we do not now know any painted glass to which an earlier date than the eleventh century can be assigned with certainty." Two specimens, and no more, of this century, are figured by M. Lasteyrie. The painted windows of the twelfth and thirteenth centuries are nearly of the same character. They consist of little historical medallions, distributed over mosaic grounds composed of coloured (not painted) glass, borrowed from preceding centuries. Fine examples from the church of St Denys and La Sainte Chapelle at Paris, of the twelfth and thirteenth centuries, are figured by M. Lasteyrie, and also by M. Labarte, who has many beautiful remarks on their harmony with the buildings to which they belong, on the elegance of their form, the richness of their details, and the brilliancy of their colours. In the fourteenth century, when examples become common, the glass-painters copied nature with more fidelity, and exchanged the violet-tinted masses, by which the flesh-tints had been rendered, for a reddish gray colour, painted upon white glass, which approached more nearly to nature. Large single figures now often occupy an entire window. The improvement in drawing and colouring is a compensation for the more striking effects of the brilliant yet mysterious examples of the preceding centuries; and the end of the fourteenth century is one of the finest epochs in the history of painted glass. Painting on glass followed the progress of painting in oils in the age which followed; and artists more and more aimed at producing individual works; and in the latter half of the fifteenth century buildings and landscapes in perspective were first introduced. The decorations which surround the figures being borrowed from the architecture of the time have often a very beautiful effect. But the large introduction of *grisailles* deprives the windows of this period of the transparent brilliancy of the coloured mosaics of the earlier glass-painting. In the sixteenth century, however, glass was nothing more than the material subservient to the glass-painter, like canvas to the oil-painter. Small pictures very highly finished were executed after the designs of Michael Angelo, Raffaelle, and the other great painters of the Renaissance. "But," as M. Labarte truly says, "the era of glass-painting was at an end. From the moment that it was attempted to transform an art of purely monumental decoration into an art of expression, its intention was perverted, and this led of necessity to its ruin. The resources of glass-painting were more limited than those of oil, with which it was unable to compete. From the end of the sixteenth century the art was in its decline, and towards the middle of the seventeenth was" almost "entirely given up." Our own age has seen its revival, and though the success has been indeed great, we may hope that the zenith has not yet been reached. "It is," says Mr Winston, "a distinct and complete branch of art, which, like many other medieval inventions, is of universal applicability, and susceptible of great improvement." I have been a little more diffuse on glass-painting than on some other subjects, as it is a purely medieval art, and one which has now acquired a living interest. Various examples of the different styles will easily suggest themselves to many, or, if not, they may be studied in the splendid work of M. Lasteyrie, entitled *Histoire de la Peinture sur Verre d'après ses monuments en France*, and on a smaller scale in Mr Winston's valuable *Hints on Glass-painting*.

With regard to the architectural monuments of the medieval world, I may, in addressing such an audience, consider them to be sufficiently well known for my present purpose, which is to give an indication, and little more, of the archæological remains which have come down to our own days. Medieval architecture is in itself a boundless subject; and as I have not specially studied it, I could not, if I would, successfully attempt an epitome of its various forms of Byzantine, Saracenic, Romanesque, Lombardic, and of infinitely diversified Gothic. For a succinct yet comprehensive view of all these and more, I must refer you to Mr Fergusson's *Handbook of Architecture*. Yet when we let our imagination idly roam over Europe, and the adjoining regions of Asia and Africa, what a host of architectural objects flits before it in endless successions of variety and beauty! Think of Justinian's Church of St Sophia, which he boasted had vanquished Solomon's temple, and again of St Mark's at Venice, as Byzantine examples. Think next of the mosque of the Sultan Hassan, and of the tombs of the Memlooks mingled with lovely minarets and domes at Cairo; of the Dome of the Rock at Jerusalem; of the Alhambra in Spain, with all the witchery of its gold and azure decorations. Float, if you will, along the banks of the Rhine or the Danube (as many of us have actually done), and conjure up the majestic cathedrals, the spacious monasteries and the ruined castles, telling of other days, with which they are fringed. Let the bare mention of the names of Milan, Venice, Rome; again of Paris, Rheims, Chartres, Amiens, Troyes, Rouen, Avignon; and in fine those of Antwerp, Louvain, and Brussels, suggest their own stories. Yet the magnificent structures, secular and ecclesiastical, which I have either named or hinted at, need not make us ashamed of our own country. We are surrounded on all sides by an archæology which is emphatically an archæology of progress, and we may justly be proud of it as Englishmen. In this University and its immediate neighbourhood we have fine specimens of Saxon, Norman, Early English, Decorated, and Perpendicular styles of Gothic architecture; and as regards the last of them, one of the most splendid examples in the world. In the opinion of competent judges the English cathedrals, while surpassed in size by many on the Continent, are in excellence of art superior to those of France or of any country in Europe. "Nothing can exceed the beauty of the crosses which Edward I. erected on the spots where the body of Queen Eleanor rested on its way to London." Some of these, Waltham for example, are quite equal to anything of their class found on the Continent. "The vault of Westminster Abbey" (says Mr Fergusson, on whose authority I make almost every statement relating to medieval architecture) "is richer and more beautiful in form than any ever constructed in France;" the triforium is as beautiful as any in existence; and its appropriateness of detail and sobriety of design render it one of the most beautiful Gothic edifices in Europe.

I thus conclude my sketch, such as it is, of the archæology of the world. Its aim has been to bring under review the rude implements and weapons of primeval man; the colossal structures of civilised man in Egypt and India; the strangely-compounded palace-sculptures of Assyria and Babylonia; the exquisitely ornamented columns of

Persian halls; the massive architecture of Phœnicia; the Gothic-like rock-tombs of Lycia; the lovely temples, and incomparable works of art of every kind, great and small, of Greece; the military impress of Roman conquest; the medieval works of art in ivory, in enamel, in glass-painting, as well as its glorious architectural remains, connecting the middle ages with our own times. It has been drawn, as I observed at the outset, under very adverse circumstances, and must on that account venture to sue for much indulgence. It is open, no doubt, to many criticisms: I expect to be charged with grievous sins of omission, and perhaps of commission also: nor do I suppose that I could entirely vindicate myself from such charges. Worse than all perhaps, I have exposed myself to the unanswerable sarcasm that I have talked about many subjects of which I know but little. If, however, I have been able to compile from trustworthy sources or manuals so much respecting those particular branches of archæology which I have not studied, as to bring before you their salient features in an intelligible manner, that is enough for my purpose. I want no more, and I pretend to no more; and I am conscious enough that even this purpose has been but feebly accomplished. Tediousness, indeed, in dealing with numerous details could hardly be altogether avoided; but this is so much lighter a fault than an indulgence in mere platitudes, running smoothly and amusingly, but emptily withal, that I shall hear your verdict of *guilty* with composure.

It now only remains that I should very briefly point out what qualifications are necessary for an archæologist, and also the pleasure and advantage which result from his pursuits.

With regard to the first of these matters, the qualifications necessary for an archæologist, they are to some considerable extent the same as are necessary for a naturalist.

Like the naturalist, the antiquary must in the first place bring together a large number of facts and objects. This is, no doubt, a matter of great labour, but believe me, '*labor ipse voluptas.*' The labour is its own ample reward. The hunting out, the securing, and the amassing facts and objects of antiquity, or of natural history, are the field-sports of the learned or scientific Nimrod. In a certain sense every archæologist *must* be a collector; he must be mentally in possession of a mass of facts and objects, brought together either by himself or by others. It is not absolutely necessary that he should be a collector, in the sense of being owner of a collection of his objects of study; in some departments indeed of archæology to amass the objects themselves is impossible: who, for instance, can collect Roman roads or Gothic cathedrals? models, plans, and drawings, are the only substitutes possible. But, with the facts relating to his favourite objects, and also as much as possible with the objects themselves, he must be familiar.

Yet this familiarity will not be enough to make him an archæologist. Such knowledge may be possessed, and very often is possessed, by a mere dealer in antiquities. The true antiquary must not only be well acquainted with his facts, but he must also, when there are sufficient data, proceed to reason upon them. He puts them together, and considers what story they have to render up. We saw a beautiful illustration of this in the joint labours of the Scandinavian antiquaries and naturalists. The order and sequence of the stone, bronze, and iron ages, were distinctly made out; and even their chronology may one day be discovered. The antiquary is enabled to form some judgment of the civilisation, the arts, and the religion of the nations whose remains he studies. Very often, as in the Roman series of coins, he makes out political events in their history, and assigns their dates. He determines the place of things in the historical series, much as the naturalist does in the natural series.

Like the naturalist also he must be a man of learning, *i. e.* he must be acquainted with what has been written by his fellow-labourers in the same branch of study. Few know, prior to experience, what a serious business this is. The bibliography of every department of archæology, as well as of natural history, is now becoming immense.

But besides a knowledge of facts, and objects, and books, there are one or two other qualifications necessary for many departments of archæology, the want of which has been very prejudicial to some distinguished writers. Exact scholarship is one of these qualifications. I do not merely mean that if a man be engaged in Greek archæology, he must be aware of the passages of Greek authors, in which the vases or the coins he is talking about are alluded to, though he must certainly be acquainted with these, and possess sufficient scholarship to construe them correctly; but he must also be able to interpret his written archæological monuments, such as his inscriptions and the legends of his coins. This is oftentimes no easy matter, and it requires a knowledge of strange words and dialects. Moreover, if an inscription or a legend be mutilated (and this is very frequently the case), unless the archæologist has an accurate knowledge of the language in which it is written, whatever that may be, Greek, Latin, Norman-French, or any other, what hope is there that he will ordinarily be able to restore it, and having so done interpret it with security or satisfaction? As one illustration of many, I will cite Prof. Ramsay's remark on Nibby's dissertation *Delle vie degli Antichi:* "In the first part of this article (on Roman roads) his essay has been closely followed. *Considerable caution, however, is necessary in using the works of this author,* who, although a profound local antiquary is by no means an accurate scholar[1]." Mr Bunbury, while pointing out the advantages which scholars would derive from

---

[1] See Smith's *Dict. Gr. and Rom. Antiq.* s. v. Viæ.

some acquaintance with archæology, points out by implication the advantage which archæologists would derive from scholarship. "In this country," says he, "the study of archæology is but too much neglected; it forms no part of the ordinary training of our classical scholars at the Universities, and is rarely taken up by them in after life. It is generally considered as the exclusive province of the professed antiquarian, who has seldom undergone that early training in accurate scholarship, which is regarded, and we think with perfect justice, by the student from Oxford or Cambridge, as the indispensable foundation of sound classical knowledge[1]." I think he is a little over-severe on us; living men like Mr C. T. Newton, Mr Waddington, Mr Vaux, Mr C. W. King, Mr C. K. Watson, and, last, but not least, like himself, to whom others might be added, prove that his assertions must be taken *cum grano;* even if it be true that this country has produced no work connected with ancient art which can be compared with the writings of Gerhard, or Welcker; of Thiersch, or Karl Otfried Müller[2].

[1] *Edinburgh Review*, u. s.

[2] I feel a little inclined to dispute this: Stuart, one of the authors of the *Antiquities of Athens*, which have been continued by other very able hands, and have also been translated into German, may, perhaps, take rank with the authors named in the text. K. O. Müller himself calls Millingen's *Ancient Unedited Monuments* (London, 1822) "a model of a work;" and though without doubt Millingen is inferior to Müller in scholarship and in acquaintance with books, he is

Another thing very desirable for the successful prosecution of some branches of archæology is an appreciation of art. Without it we cannot judge of the value of many antiques, or enter into their spirit or feeling; we neither discern their excellencies nor their deficiencies. Mr King, who has made the province of ancient gems peculiarly his own, justly calls them "little monuments of perfect taste,...only to be appreciated by the educated and practised eye[1]." Moreover, this is the very knowledge often so requisite for distinguishing genuine antiquities from modern counterfeits. The modern forgers, who fabricate Greek coins from false dies, do not often reach the freedom and beauty of the originals; though it must be confessed that some of them, as Becker, have carried their execrable art to a very high perfection. It is but rarely that those men meet with the punishment they deserve; yet it is satisfactory to know that Charles Patin, great scholar and great antiquary as he was, was banished by Lewis XIV. from his court for ever, for selling him a false coin of Otho; and that a manufacturer of antiques in the East, near Bagdad I believe, lately received by order of the Turkish governor a sound bastinado on the soles of his feet for reproducing the idols of misbelievers of old time.

probably at least his equal as a practical archæologist. Colonel Leake's *Numismata Hellenica* (London, 1856) may also be cited as an admirable combination of learning with practical archæology.

[1] *Antique Gems*, Introd. p. xxiii. London, 1860.

A knowledge of natural history in fine is occasionally very useful to an antiquary. I will give two instances, not at all generally known, one taken from zoology, one from botany. On the reverse of the splendid Greek coins of Agrigentum a crab is commonly represented. To an ignorant eye the crab looks much like the crab in our shops here in Cambridge; the zoologist recognises in it the fresh-water crab of the regions of the Mediterranean; the numismatist, profiting by this knowledge, sees at once that the type of the coin symbolizes not the harbour of Agrigentum, as he had supposed, but its river. Again, on the reverse of the beautiful Greek coins of Rhodes occurs a flower, about which numismatists have disputed since the time of Spanheim, whether it was the flower of the rose or of the pomegranate. Even Col. Leake has here taken the wrong side, and decided in favour of the pomegranate; the divided calyx at once shews every botanist that the representation is intended for the rose, conventional as that representation may be, from which flower the island derives its name.

These are, I think, the principal qualifications which are necessary or desirable for the archæologist. It only remains that I should point out briefly some of the pleasures and advantages that result from his pursuits. For I shall not so insult any one of you, who are here present, as to suppose that this question is lurking secretly in your mind, "Is there any good in archæology at all?

To what practical end do your researches tend?" My learned predecessor well says that "this question is sometimes put to the lover of science or letters by those from whom nature has withheld the faculty of deriving pleasure from the exercise of the intellect, and he feels for the moment degraded to the level of such." It is not so clear however that the fault must be put to the account of nature. Rather, we may say,

Homine *imperito* nunquam quidquam injustius,
Qui nisi quod ipse facit, nihil rectum putat.

"No one," says a Swedish scholar of the seventeenth century, "blames the study of antiquity without evidencing his own ignorance; as they that esteem it do credit to their own judgment; so that to sum up its advantages we may assert, there is nothing useful in literature, if the knowledge of antiquity be judged unprofitable[1]." It is doubtless one of the many charms of archæology that it illustrates and is illustrated by literature; indeed, some knowledge of antiquity is little less than necessary for every man of letters. Unless we have some knowledge of the objects whose names occur in ancient literature, we lose half the pleasure of reading it. In reading the New Testament, I can certainly say for myself, that I derive more pleasure from the narrative of the woman who poured the contents of the alabaster box

[1] Figrelius, quoted in the *Museum of Classical Antiquities*, Vol. I. p. 4.

over the head of Jesus, now that I know what an *alabastron* is, and how its contents would be extracted; and in the same way I appreciate the remark made by the silversmith in the Acts, that all Asia and the world worshipped the Ephesian Diana, now that I know her image to be stamped not on the coins of Ephesus only, but on many other cities throughout Asia also. Here, I think, we have pleasure and profit combined in one. Instances are abundant where monuments illustrate profane authors. The reader of Aristophanes will be pleased to recognise among the earliest figures on vases that of the ἱπποαλεκτρύων, the cock-horse, or horse-cock, which cost Bacchus a sleepless night to conceive what manner of fowl it might be. "The Homeric scholar again," it has been said, "must contemplate with interest the ancient pictures of Trojan scenes on the vases, and can hardly fail to derive some assistance in picturing them to his own imagination, by seeing how they were reproduced in that of the Greeks themselves in the days of Æschylus and Pindar[1]."

Further, not only is ancient literature, but also modern art, aided by archæology. It is well known how, in the early part of the thirteenth century, Niccola Pisano was so attracted by a bas-relief of Meleager, which had been lying in Pisa for ages unheeded, "that it became the basis of his studies and the germ of true taste in Italy."

[1] *Edinburgh Review*, u. s.

In the Academy of St Luke at Rome, and in the schools established shortly afterwards at Florence by Lorenzo de' Medici, the professors were required to point out to the students the beauty and excellence of the works of ancient art, before they were allowed to exercise their own skill and imagination. Under the fostering patronage of this illustrious man and of his not less illustrious son a galaxy of great artists lighted up all Europe with their splendour. Leon Batista Alberti, one of the greatest men of his age, and especially great in architecture, was most influential in bringing back his countrymen to the study of the monuments of antiquity. He travelled to explore such as were then known, and tells us that he shed tears on beholding the state of desolation in which many of them lay. The prince of painters, Raffaelle,

> timuit quo sospite vinci
> Rerum magna parens et moriente mori,

and the prince of sculptors, Michael Angelo, both drew their inspiration from the contemplation of the art-works of antiquity. The former was led to improve the art of painting by the frescoes of the baths of Titus, the latter by the sight of a mere torso imbibed the principles of proportion and effect which were so admirably developed in that fragment[1]. And not only the arts of sculpture

---

[1] For this and the preceding facts see the *Museum of Classical Antiquities*, Vol. I. pp. 13—15. The frescoes of the baths

and painting, but those which enter into our daily life, are furthered by the wise consideration of the past. Who can have witnessed the noble exhibitions in Hyde Park or at Kensington without feeling how much the objects displayed were indebted to Hellenic art? In reference to the former of these Mr Wornum says: "Repudiate the idea of copying as we will, all our vagaries end in a recurrence to Greek shapes; all the most beautiful forms in the Exhibition, (whether in silver, in bronze, in earthenware, or in glass,) are Greek shapes; it is true often disfigured by the accessory decorations of the modern styles, but still Greek in their essential form[1]."

And yet I must, in concluding this Introductory Lecture, most strongly recommend to you the study of archæology, not only for its illustration of ancient literature, not only for its furtherance of modern art, but also, and even principally, for its own sake. "Hæc studia adolescentiam alunt, senectutem oblectant, secundas res ornant, adversis perfugium ac solatium præbent; delectant domi, non impediunt foris, pernoctant nobiscum, peregrinantur, rusticantur[2]." Every one who follows a pursuit in addition to the routine duties of life has,

of Titus have subsequently lost their brilliancy. See Quatremère de Quincy's *Life of Raphael,* p. 263. Hazlitt's Translation. (Bogue's European Library).

[1] *The Exhibition as a Lesson in Taste,* p. xvii. * * * (Printed at the end of the *Art-Journal Illustrated Catalogue,* 1851).

[2] Cicero *pro Archia poeta,* c. vii.

by so doing, a happiness and an advantage of which others know little. The more elevated the pursuit, the more exquisite the happiness and the more solid the advantage. Now if

> The proper study of mankind is man,

then most assuredly archæology is one of the most proper pursuits which man can follow. For she is the interpreter of the remains which man in former ages has left behind him. By her we read his history, his arts, his civilisation; by her magical charms the past rises up again and becomes a present; the tide of time flows back with us in imagination; the power of association transports us from place to place, from age to age, suddenly and in a moment. Again the glories of the nations of the old world shine forth;

> Again their godlike heroes rise to view,
> And all their faded garlands bloom anew.

To adopt and adapt the words of one who is both a learned archæologist and a learned astronomer of this University, I feel that I may, under any and all circumstances, impress upon your minds the utility and pleasure of "every species and every degree of archæological enquiry." For "history must be looked upon as the great instructive school in the philosophical regulation of human conduct," as well as the teacher "of moral precepts" for all ages to come; and no "better aid can be appealed to for" the discovery, for "the confirmation, and for the demonstration of the facts of history, than the energetic pursuit of archæology"[1].

[1] See an address delivered at an Archæological meeting at Leicester, by John Lee, Esq., LL.D. (*Journal of Archæol. Association* for 1863, p. 37).

NOTES.

Pp. 15—20. Nearly everything contained in the text relating to prehistoric Europe will be found in the *Revue Archéologique* for 1864, and in Sir C. Lyell's *Antiquity of Man*, London, 1863; see also for Thetford, *Antiq. Commun.* Vol. I. pp. 339—341, (Cambr. Antiq. Soc. 1859); but the following recent works (as I learn from Mr Bonney, who is very familiar with this class of antiquities) will also be found useful to the student:

Prehistoric Times. By John Lubbock, F.R.S. London, 1865. 8vo.

The Primeval Antiquities of Denmark. By Prof. Worsäae. London, 1849. 8vo. (Engl. Transl.).

Les Habitations Lacustres. Par F. Troyon. Lausanne, 1860.

Les Constructions Lacustres du Lac de Neufchâtel. Par E. Desor. Neufchâtel, 1864.

Antiquités Celtiques et Antédiluviennes. Par Boucher de Perthes. Paris, 1847.

Die Pfahlbauten. Von Dr Ferd. Keller. Ber. I—V. (*Mittheilungen der Antiquarischen Gesellschaft in Zurich*). 1854, sqq. 4to.

Die Pfahlbauten in den Schweizer-Seen. Von L. Staub. Zurich, 1864. 8vo.

Besides these there are several valuable papers in the *Transactions of the Royal, Geological, and Antiquarian Societies* (by Messrs John Evans, Prestwich, and others), the *Natural History Review*, and other Periodicals.

p. 26. For the literature relating to ancient Egypt see Mr R. S. Poole's article on Egypt, in Smith's *Dictionary of the Bible*, Vol. I. p. 512.

pp. 29—31. Besides the works of Robinson, De Saulcy, Lewin, Thrupp, and others, the following books may be mentioned as more especially devoted to the archæology of Jerusalem:

The Holy City. By George Williams, B.D. (Second edition, including an architectural History of the Church of the Holy Sepulchre by the Rev. Robert Willis, M.A., F.R.S. 1849.)

Jerusalem Explored. By Ermete Pierotti. Translated by T. G. Bonney, M.A. 1864.

Le Temple de Jérusalem. Par le Comte Melchior de Vogüé, 1865. The Count considers none of the present remains of the Temple to be earlier than the time of Herod.

To these I should add Mr Williams' and Mr Bonney's tracts, directed against the views of Mr Fergusson, in justification of those of Dr Pierotti.

p. 31, l. 20. From some remarks made to me by my learned friend, Count de Vogüé, I fear that this is not so certain a characteristic of Phœnician architecture as has been commonly supposed. He assigns some of the bevelled stones which occur in Phœnicia to the age of the Crusades.

p. 31, last line. For the very remarkable Phœnician sarcophagus discovered in 1855, and for various references to authorities on Phœnician antiquities, see Smith's *Dict. of the Bible*, Vol. II. p. 868, and Vol. III. p. 1850.

p. 36. As a general work on Greek and Roman Coins Eckhel's *Doctrina Numorum Veterum* (Vindobonæ, 1792—1828, with Steinbuchel's *Addenda*, 8 Vols. 4to.) still remains the standard, though now getting a little out of date.

The same remark must be made of Mionnet's great work, *Description de Médailles Antiques, Grecques et Romaines*, Paris, 1806—1813 (7 Vols.), with a supplement of 9 Vols. Paris, 1819—1837, giving a very useful *Bibliothèque Numismatique* at the end; to which must be added his *Poids des Médailles Grecques*, Paris, 1830. These seventeen volumes comprise the

Greek coins: the other part of his work, *De la Rareté et du Prix des Médailles Romaines*, Paris, 1827, in two volumes, is now superseded.

Since Mionnet's time certain departments of Greek and other ancient numismatics have been much more fully worked out, especially by the following authors:

De Luynes (coins of Satraps; also of Cyprus); L. Müller (coins of Philip and Alexander; of Lysimachus; also of Ancient Africa); Pinder (Cistophori); Beulé (Athenian coins); Lindsay (Parthian coins); Longpérier, and more recently Mordtmann (coins of the Sassanidæ); Carelli's plates described by Cavedoni (coins of Magna Græcia, &c.); other works of Cavedoni (Various coins); Friedländer (Oscan coins); Sambon (coins of South Italy); De Saulcy, Levy, Maddon (Jewish coins); V. Langlois (Armenian, also early Arabian coins); J. L. Warren (Greek Federal coins; also more recently, copper coins of Achæan League); R. S. Poole (coins of the Ptolemies); Waddington (Unedited coins of Asia Minor).

For Roman and Byzantine coins (including Æs grave and Contorniates) see the works of Marchi and Tessieri, Cohen, Sabatier, and De Saulcy.

Others, as Prokesch-Osten, Leake, Smyth, Hobler, and Fox, have published their collections or the unedited coins of them; and all the numismatic periodicals contain various previously unedited Greek and Roman and other ancient coins.

p. 40. Fabretti's work is entitled, *Glossarium Italicum in quo omnia vocabula continentur ex Umbricis, Sabinis, Oscis, Volscis, Etruscis, ceterisque monumentis collecta, et cum interpretationibus variorum explicantur* (Turin, 1858—1804). Many figures of the antiquities, on which the words occur, are given in their places.

p. 43. Cromlechs in some, if not in all cases, appear to be the skeletons of barrows.

p. 44. The following works will be found useful for the student of early British antiquities:

Pictorial History of England, Vol. I. Lond. 1838.

Archæological Index to remains of Antiquity of the Celtic, Romano-British, and Anglo-Saxon periods. By J. Y. Akerman, F.S.A. London, 1847 (with a classified index of the Papers in the *Archæologia*, Vols. I—XXXI).

Ten years' diggings in Celtic and Saxon Grave Hills in the Counties of Derby, Stafford, and York, from 1848—1858. By Thomas Bateman. London, 1861. A most useful work, which will indicate the existence of many others. In connection with this see Dr Thurnam's paper on British and Gaulish skulls in *Memoirs of Anthropological Soc.* Vol. I. p. 120.

The Land's End District, its Antiquities, Natural History, &c. By Richard Edmonds. London, 1862.

Catalogue of the Antiquities of Stone, Earthen, and Vegetable Materials, in the Museum of the Royal Irish Academy. By W. R. Wilde, M.R.I.A. Dublin, 1857.

The Coins of the Ancient Britons. By John Evans, F.S.A. The plates by F. W. Fairholt, F.S.A. London, 1864. By far the best and most complete work hitherto published on the subject.

Also, the *Transactions* of various learned Societies in Great Britain and Ireland, among which the *Archæologia Cambrensis* is deserving of special mention.

For the Romano-British Antiquities may be added Horsley's *Britannia Romana*, 1732; Roy's *Military Antiquities of the Romans in Britain*, 1793; Lysons' *Relliquiæ Britannico-Romanæ*. London, 1813, 4 Vols. fol.

Monographs on York, by Mr Wellbeloved; on Richborough and other towns, by Mr C. R. Smith; on Aldborough, by Mr H. E. Smith; on Wroxeter, by Mr Wright; on Caerleon, by Mr Lee; on Cirencester, by Messrs Buckman and Newmarch; on Hadrian's wall, by Dr Bruce; on various excavations in Cambridgeshire, by the Hon. R. C. Neville.

p. 45. For the Roman Roads, &c. in Cambridgeshire, see Prof. Charles C. Babington's *Ancient Cambridgeshire*, Cambr. 1853 (Cambr. Ant. Soc.).

— No doubt need have been expressed about Wroxeter, which should hardly have been called 'our little Pompeii'; the area of Wroxeter being greater, however less considerable the remains. See Wright's *Guide to Uriconium*, p. 88. Shrewsbury, 1860. For various examples of Roman wall-painting in Britain see *Reliq. Isur.* by H. E. Smith, p. 18, 1852.

p. 46. For Romano-British coins see

Coins of the Romans relating to Britain, described and illustrated. By J. Y. Akerman, F.S.A. London, 1844.

Petrie's *Monumenta Historica Britannica*, Pl. I—XVII. London, 1848 (for beautiful figures).

Others, published by Mr C. R. Smith in his valuable *Collectanea Antiqua;* also by Mr Hobler, in his *Records of Roman History, exhibited on Coins.* London, 1860. Others in the *Numismatic Chronicle*, in the *Transactions of the Cambridge Antiquarian Society*, and perhaps elsewhere.

For medieval and modern numismatics in general we may soon, I trust, have a valuable manual (the MS. of which I have seen) from the pen of my learned friend, the Rev. W. G. Searle. He has favoured me with the following notes:

On medieval and modern coins generally we have

Appel, *Repertorium sur Münzkunde des Mittelalters und der neuern Zeit*, 6 Vols. 8vo. Peath, 1820—1829.

Barthélémy, *Manuel de Numismatique du moyen âge et moderne.* Paris, 1851. 12mo.

The bibliography up to 1840 we get in

Lipsius, *Biblioth. Numaria*, Leipz. 1801 (2 Vols.) 8vo., and in Leitzmann, *Verzeichniss aller seit 1800 erschienenen Numism. Werke*, Weissensee, 1841, 8vo.

On medieval coins, their types and geography, we have

J. Lelewel, *La Numismatique du Moyen-âge, considérée sous le rapport du type.* Paris, 1835, 2 vols. 8vo. Atlas 4to.

Then there are the great Numismatic Periodicals:

Revue Numism. 8vo. Paris, 1836.

Revue de la Num. Belge, 8vo. Brussels, 1841.

Leitzmann, *Numismatische Zeitung*, 4to. Weissensee, 1834.

On Bracteates:

Mader, *Versuch über die Bracteaten.* Prague, 1797, 4to.

And the great Coin Catalogues of

Welzl v. Wellenheim. 3 vols. 8vo. Vienna, 1844 ff. (c. 40,000 coins).

v. Reichel at St Petersburgh, in at least 9 parts.

On current coins we have

Lud. Fort, *Neueste Münzkunde*, engravings and descr. 8vo. Leipzig, 1851 ff.

p. 45. For almost everything relating to ivories and for a great deal on the subjects which follow, see *Handbook of the Arts of the Middle Ages and Renaissance*, Translated from the French of M. Jules Labarte, with notes, and copiously illustrated, London, 1855, which will lead the student to the great authorities for medieval art, as Du Sommerard, &c. I have also examined and freely used *Histoire des Arts industriels au moyen âge et à l'époque de la Renaissance*, Par Jules Labarte. Paris, 1864, 8vo. 2 volumes; accompanied by an album in quarto with descriptions of the plates, also in two volumes.

p. 47. For examples of medieval calligraphy and illuminations see Mr Westwood's *Palæographia Sacra Pictoria*, (Lond. 1845), and his *Illuminated Illustrations of the Bible*, (London, 1846).

p. 48. A good deal of information about Celtic, Romano-British, and medieval pottery will be found in Mr Jewitt's *Life of Wedgwood*, London, 1865. For ancient pottery in general (excluding however the medieval) see Dr Birch's *Ancient Pottery and Porcelain*, London, 1858, which will conduct the student to the most authentic sources of information. In connection with this should be studied Mr Bunbury's article in the *Edinburgh Review* for 1858, to which Mr Oldfield's paper on Sir W. Temple's vases in the *Transactions of the Royal Soc. of Lit.* Vol. VI. pp. 130—149 (1859), may be added.

— For medieval sculpture see Flaxman's *Lectures*. The 'horrible and burlesque' style of the earlier ages was discarded in the thirteenth century, when the art revived in Italy. Italian artists executed various sepulchral statues in this country, which possess considerable merit, as do others by native artists, but the great beauty of our sepulchral monuments consists in their architectural decorations.

p. 49. For the coinage of the British Islands see the works of Ruding, Hawkins, and Lindsay, also for the Saxon coins found in great numbers in Scandinavia, Hildebrand and Schröder. Humphreys' popular work on the coinage of the British Empire, so far as the plates are concerned, is useful, but the author is deficient in scholarship.

p. 52. For the statements here made on oil-painting see Bryan's *Dict. of Painters and Engravers*, by Stanley, (London, 1849), under Van Eyck, and Sir C. L. Eastlake's *Materials for a History of Oil-painting*. (London 1847.)

p. 53. For medieval brasses, see

Bowtell, *Monumental Brasses and Slabs*. London, 1847, 8vo.

———— *Monumental Brasses of England, a Series of engravings in wood.* London, 1849.

Haines, *Manual of Monumental Brasses.* 2 parts. London, 1861, 8vo. This contains also a list of all the brasses known to him as existing in the British Isles. Mr Way has given an account of foreign sepulchral brasses in *Archæol. Journ.*, Vol. VII.

p. 56. Several English frescoes are described and figured in the *Journal of the Archæological Association,* passim.

p. 62, l. 13. The omission of ancient costume has been pointed out to me. The *actually existing* specimens however are mostly very late; with the exception of a few articles of dress found in Danish sepulchres of the bronze period, or in Irish peat bogs of uncertain date, the episcopal vestments of Becket now preserved at Sens are the earliest which occur to my recollection; and there are few articles of dress, I believe, so early as these. However both ancient and medieval costume is well known from the *representations* on monuments of various kinds. See *inter alia* Hope's *Costume of the Ancients;* Becker's *Gallus* and *Charicles;* Strutt's *Dress of the English People,* edited by Planché, (Lond. 1842); Shaw's *Dresses and Decorations of the Middle Ages.*

p. 67. The statement about Patin is made on the authority of a note in Warton's edition of Pope's Works, Vol. III. p. 306. (London 1797.)

p. 68. The remark about the crab was made to me by the late Mr Burgon, and I do not know whether it has ever been printed; its truth seems pretty certain. For the Rhodian symbol see my paper in the *Numismatic Chronicle* for 1864, pp. 1—6.

PREPARING FOR PUBLICATION,

An Introduction to the Study of Greek Fictile Vases; their Classification, Subjects, and Nomenclature. Being the substance of the Disney Professor's Lectures for 1865, and of those which he purposes to deliver in 1866.

CAMBRIDGE.
January, 1866

LIST OF WORKS

PUBLISHED BY

DEIGHTON, BELL, AND CO.

Agents to the University.

———— ALFORD's (DEAN) Greek Testament: with a critically revised Text; a Digest of Various Readings; Marginal References to Verbal and Idiomatic Usage; Prolegomena; and a Critical and Exegetical Commentary. For the use of Theological Students and Ministers. By HENRY ALFORD, D.D., Dean of Canterbury.

    Vol. I. *Fifth Edition*, containing the Four Gospels. 1l. 8s.

    Vol. II. *Fifth Edition*, containing the Acts of the Apostles, the Epistles to the Romans and Corinthians. 1l. 4s.

    Vol. III. *Fourth Edition*, containing the Epistle to the Galatians, Ephesians, Philippians, Colossians, Thessalonians,—to Timotheus, Titus, and Philemon. 18s.

    Vol. IV. Part I. *Third Edition*, containing the Epistle to the Hebrews, and the Catholic Epistles of St. James and St. Peter. 18s.

    Vol. IV. Part II. *Third Edition*, containing the Epistles of St. John and St. Jude, and the Revelation. 14s.

———— New Testament for English Readers. Containing the Authorised Version, with additional corrections of Readings and Renderings; Marginal References; and a Critical and Explanatory Commentary. By HENRY ALFORD, D.D., Dean of Canterbury. In two volumes.

    Vol. I. Part I. Containing the First Three Gospels. 12s.

    Vol. I. Part II. Containing St. John and the Acts. 10s. 6d.

    Vol. II. Part I. Containing the Epistles of St. Paul. 16s.

———— A Plea for the Queen's English; Stray Notes on Speaking and Spelling. By HENRY ALFORD, D.D. Small 8vo. 5s.

———— Letters from Abroad. By HENRY ALFORD, D.D. Small 8vo. 7s. 6d.

———— Family Prayers for the Christian Year. By HENRY ALFORD, D.D. Small 8vo.   (*Preparing.*)

APOSTOLIC EPISTLES, A General Introduction to the, with a Table of St. Paul's Travels, and an Essay on the State after Death. Second Edition, enlarged. To which are added, A Few Words on the Athanasian Creed, on Justification by Faith, and on the Ninth and Seventeenth Articles of the Church of England. By a Bishop's Chaplain. 8vo. 5s. 6d.

BABINGTON's (Churchill, B.D., F.L.S.) An Introductory Lecture on Archæology, delivered before the University of Cambridge. By Churchill Babington, B.D., F.L.S. 8vo. price 2s.

BARRETT's (A. C.) Companion to the New Testament. Designed for the use of Theological Students and the Upper Forms in Schools. By A. C. Barrett, M.A., Caius College. Fcap. 8vo. 5s.

BEAMONT's (W. J.) Cairo to Sinai and Sinai to Cairo. Being an Account of a Journey in the Desert of Arabia, November and December, 1860. By W. J. Beamont, M.A., Fellow of Trinity College, Cambridge, and Incumbent of St. Michael's, Cambridge, sometime Principal of the English College, Jerusalem. With Maps and Illustrations. Fcap. 8vo. 5s.

———— A Concise Grammar of the Arabic Language. Revised by Sheikh Ali Nady El Barrany. By W. J. Beamont, M.A. Price 7s.

BLUNT's (Rev. J. J.) Five Sermons preached before the University of Cambridge. The first Four in November, 1851, the Fifth on Thursday, March 5th, 1849, being the Hundred and Fiftieth Anniversary of the Society for Promoting Christian Knowledge. By the late Rev. J. J. Blunt, B.D., Lady Margaret Professor of Divinity. 8vo. 5s. 6d.

CONTENTS: 1. Tests of the Truth of Revelation.—2. On Unfaithfulness to the Reformation.—3. On the Union of Church and State.—4. An Apology for the Prayer-Book.—5. Means and Method of National Reform.

BROWNE's (Bp.) Messiah as Foretold and Expected. A Course of Sermons relating to the Messiah, as interpreted before the Coming of Christ. Preached before the University of Cambridge in the months of February and March, 1862. By the Right Reverend E. Harold Browne, D.D., Bishop of Ely. 8vo. 4s.

CAMBRIDGE University Calendar, 1865. 6s. 6d.

CAMPION's (Rev. W. M.) Nature and Grace. Sermons preached in the Chapel Royal, Whitehall, in the year 1863-3-4. By William Magan Campion, B.D., Fellow and Tutor of Queens' College, Cambridge, Rector of St. Botolph's, Cambridge, and one of her Majesty's Preachers at Whitehall. Small 8vo. 6s. 6d.

CHEVALLIER's (T.) Translation of the Epistles of Clement of Rome, Polycarp and Ignatius; and of the Apologies of Justin Martyr and Tertullian; with an Introduction and Brief Notes Illustrative of the Ecclesiastical History of the First Two Centuries. By T. Chevallier, B.D. *Second Edition.* 8vo. 12s.

COOPER's (C. H. and Thompson) Athenae Cantabrigienses. By C. H. Cooper, F.S.A., and Thompson Cooper, F.S.A.

This work, in illustration of the biography of notable and eminent men who have been members of the University of Cambridge, comprehends, notices of: 1. Authors. 2. Cardinals, archbishops, bishops, abbats, heads of religious houses and other Church dignitaries. 3. Statesmen, diplomatists, military and naval commanders. 4. Judges and eminent practitioners of the civil or common law. 5. Sufferers for religious and political opinions. 6. Persons distinguished for success in tuition. 7. Eminent physicians and medical practitioners. 8. Artists, musicians, and heralds. 9. Heads of colleges, professors, and principal officers of the university. 10. Benefactors to the university and colleges or to the public at large.

Volume I. 1500—1585. 8vo. cloth. 18s. Volume II. 1586—1609. 18s.
Volume III. *In the Press.*

DINGLE's (Rev. J.) Harmony of Revelation and Science. A Series of Essays on Theological Questions of the Day. By the Rev. J. Dingle, M.A., F.A.S.L., Incumbent of Lanchester, Durham. Crown 8vo. 6s.

DONALDSON's (Rev. J. W.) The Theatre of the Greeks. A Treatise on the History and Exhibition of the Greek Drama; with various Supplements. By the Rev. J. W. Donaldson, D.D. *Seventh Edition*, revised, enlarged, and in part remodelled; with numerous Illustrations from the best ancient authorities. 8vo. 14s.

———————— Classical Scholarship and Classical Learning considered with especial reference to Competitive Tests and University Teaching. A Practical Essay on Liberal Education. By the Rev. J. W. Donaldson, D.D. Crown 8vo. 5s.

D'ORSEY's (Alexander J. D.) Study of the English Language an Essential part of a University Course: An Extension of a Lecture delivered to the Royal Institution of Great Britain, February 1, 1861. With Coloured Language-Maps of the British Isles and Europe. By Alexander J. D. D'Orsey, B.D., English Lecturer at Corpus Christi College, Cambridge, &c. Crown 8vo. cloth. 2s. 6d.

ELLIS' (Rev. A. A.) Bentleii Critica Sacra. Notes on the Greek and Latin Text of the New Testament, extracted from the Bentley MSS. in Trinity College Library. With the Abbé Rulotta's Collation of the Vatican MS., a specimen of Bentley's intended Edition, and an account of all his Collations. Edited, with the permission of the Masters and Seniors, by the Rev A. A. Ellis, M.A., late Fellow of Trinity College, Cambridge. 8vo. 8s. 6d.

LIST OF WORKS PUBLISHED BY

ELLIS (ROBERT LESLIE). The Mathematical and other Writings of ROBERT LESLIE ELLIS, M.A., Fellow of Trinity College, Cambridge. Edited by WILLIAM WALTON, M.A., Trinity College, with a Biographical Memoir by the Very Reverend HARVEY GOODWIN, D.D., Dean of Ely. Portrait. 8vo. 16s.

EWALD's (H.) Life of Jesus Christ. By H. EWALD. Edited by OCTAVIUS GLOVER, B.D. Crown 8vo. 6s.

FRANCIS (REV. JOHN). "The Exercise of the Active Virtues, such as Courage and Patriotism, is entirely consistent with the Spirit of the Gospel;" being the Burney Prize Essay for 1863. By the Rev. JOHN FRANCIS, B.A., Vice-Principal of Bishop Otter's Training College, Chichester. 8vo. 2s.

FULLER'S (REV. J. M.) Essay on the Authenticity of the Book of Daniel. By the Rev. J. M. FULLER, M.A., St. John's College, Cambridge. 8vo. 6s.

FURIOSO, or, Passages from the Life of LUDWIG VON BEETHOVEN. From the German. Small 8vo. 4s.

GOODWIN'S (DEAN) Doctrines and Difficulties of the Christian Religion contemplated from the Standing-point afforded by the Catholic Doctrine of the Being of our Lord Jesus Christ. Being the Hulsean Lectures for the year 1855. By H. GOODWIN, D.D. 8vo. 9s.

——————— 'The Glory of the Only Begotten of the Father seen in the Manhood of Christ.' Being the Hulsean Lectures for the year 1856. By H. GOODWIN, D.D. 8vo. 7s. 6d.

——————— Parish Sermons. By H. GOODWIN, D.D. 1st Series. Third Edition. 12mo. 6s.

——————— 2nd Series. Third Edition. 12mo. 6s.

——————— 3rd Series. Second Edition. 12mo. 7s.

——————— 4th Series, 12mo. 7s.

——————— 5th Series. With Preface on Sermons and Sermon Writing. 7s.

GOODWIN's (DEAN) Four Sermons preached before the University of Cambridge, in the Season of Advent, 1858. By H. Goodwin, D.D. 12mo. 3s. 6d.

——— Christ in the Wilderness. Four Sermons preached before the University of Cambridge in the month of February, 1855. By H. Goodwin, D.D. 12mo. 4s.

——— Short Sermons at the Celebration of the Lord's Supper. By H. Goodwin, D.D. New Edition. 12mo. 4s.

——— Lectures upon the Church Catechism. By H. Goodwin, D.D. 12mo. 4s.

——— A Guide to the Parish Church. By Harvey Goodwin, D.D. Price 1s. sewed; 1s. 6d. cloth.

——— Confirmation Day. Being a Book of Instruction for Young Persons how they ought to spend that solemn day, on which they renew the Vows of their Baptism, and are confirmed by the Bishop with prayer and the laying on of hands. By H. Goodwin, D.D. Eighth Thousand. 2d., or 25 for 3s. 6d.

——— Plain Thoughts concerning the meaning of Holy Baptism. By H. Goodwin, D.D. Second Edition. 2d., or 25 for 3s. 6d.

——— The Worthy Communicant; or, 'Who may come to the Supper of the Lord?' By H. Goodwin, D.D. Second Edition. 2d., or 25 for 3s. 6d.

——— The Doom of Sin, and the Inspiration of the Bible. Two Sermons preached in Ely Cathedral; with some Prefatory Remarks upon the Oxford Declaration. By Harvey Goodwin, D.D. Fcap. 8vo. 1s. 6d.

——— Hands, Head, and Heart; or the Christian Religion regarded Practically, Intellectually, and Devotionally. In Three Sermons preached before the University of Cambridge. By Harvey Goodwin, D.D. Fcap. 8vo. 2s. 6d.

——— The Ministry of Christ in the Church of England. Four Sermons Preached before the University of Cambridge. I.—The Minister called. II.—The Minister as Prophet. III.—The Minister as Priest. IV.—The Minister Tried and Comforted. By H. Goodwin, D.D., Dean of Ely. Fcap 8vo. 2s. 6d.

GOODWIN's (Dean) The Appearing of Jesus Christ. A short Treatise by Symon Patrick, D.D., formerly Lord Bishop of Ely, now published for the first time from the Original MS. Edited by the Dean of Ely. 18mo. 5s

——— Commentaries on the Gospels, intended for the English Reader, and adapted either for Domestic or Private Use. By Harvey Goodwin, D.D. Crown 8vo.
   3. MATTHEW, 12s.   5. MARK, 7s. 6d.   8. LUKE, 9s.

——— On the Imitation of Christ. A New Translation. By the Dean of Ely. Second Edition. 18mo. 3s. 6d.
Fcap. 8vo. An Edition printed on large paper, 5s.

GREGORY (Duncan Farquharson). The Mathematical Writings of Duncan Farquharson Gregory, M.A., late Fellow of Trinity College, Cambridge. Edited by William Walton, M.A., Trinity College, Cambridge. With a Biographical Memoir by Robert Leslie Ellis, M.A., late Fellow of Trinity College. 1 vol. 8vo. 12s.

GROTE's (Rev. J.) Exploratio Philosophicæ. Rough Notes on Modern Intellectual Science. Part 1. By J. Grote, B.D., Professor of Moral Philosophy. 8vo. 5s.

HARDWICK's (Archdeacon) History of the Articles of Religion. To which is added a series of Documents from A.D. 1536 to A.D. 1615. Together with Illustrations from contemporary sources. By Charles Hardwick, B.D., late Archdeacon of Ely. *Second Edition, corrected and enlarged.* 8vo. 12s.

HUMPHRY's (Rev. W. G.) Historical and Explanatory Treatise on the Book of Common Prayer. By W. G. Humphry, B.D., late Fellow of Trinity College, Cambridge. *Third Edition, revised and enlarged.* Small post 8vo. 4s. 6d.

KENT's Commentary on International Law. Revised, with notes and Cases brought down to the present year. Edited by J. T. Abdy, LL.D., Regius Professor of Law in the University of Cambridge.
(*In the Press.*)

LAMB (Rev. John). The Seven Words Spoken Against the Lord Jesus: or, an Investigation of the Motives which led His Contemporaries to reject Him. Being the Hulsean Lectures for the year 1850. By John Lamb, M.A., Senior Fellow of Gonville and Caius College, and Minister of S. Edward's, Cambridge. 8vo. 5s. 6d.

LEAKE's (COLONEL). The Topography of Athens, with some
remarks on its Antiquity. *Second Edition.* Two vols. 8vo., with
Eleven Plates and Maps. By COLONEL LEAKE, Vice-President of the
Royal Society of Literature, of the Royal Geographical Society, &c.
(Pub. at 60s.) 18s.

——— Travels in Northern Greece. Four vols., 8vo. with
Ten Maps and Forty-four Plates of Inscriptions. By COLONEL LEAKE.
(Pub. at 60s.) 60s.

——— Peloponnesiaca, a Supplement to Travels in the
Morea. Five Maps. By COLONEL LEAKE. (Pub. at 15s.) 7s. 6d.

——— On Some Disputed Questions in Geography, with
a Map of Africa. By COLONEL LEAKE. (Pub. at 6s. 6d.) 4s. 6d.

——— Numismata Hellenica, with Supplement and Appen-
dix, completing a Descriptive Catalogue of Twelve Thousand Greek
Coins, with Notes Geographical and Historical, Map, and Index. By
COLONEL LEAKE. 4to. (Pub. at 63s.) 42s.

LEAPINGWELL's (DR. G.) Manual of the Roman Civil Law,
arranged according to the Syllabus of Dr. HALLIFAX. By G. LEAPING-
WELL, LL.D. Designed for the use of Students in the Universities and
Inns of Court. 8vo. 11s.

LEATHES (STANLEY). The Birthday of Christ, its Preparation,
Message, and Witness. Three Sermons preached before the University
of Cambridge, in December, 1865. By STANLEY LEATHES, M.A.,
Preacher and Assistant Minister, St. James's, Piccadilly, Professor of
Hebrew, King's College, London. Fcap. 8vo.

LIVINGSTONE's (DR.) Cambridge Lectures. With a Pre-
fatory Letter by the Rev. Professor SEDGWICK, M.A., F.R.S., &c.,
Vice-Master of Trinity College, Cambridge. Edited, with Introduction,
Life of Dr. Livingstone, Notes and Appendix, by the Rev. W. MONK,
M.A., F.R.A.S., &c., of St. John's College, Cambridge. With a Portrait
and Map, also a larger Map, by Arrowsmith, granted especially for
this work by the President and Council of the Royal Geographical
Society of London. Crown 8vo. 6s. 6d.

MACKENZIE (BISHOP), Memoir of the late. By the DEAN
OF ELY. With Maps, Illustrations, and an Engraved Portrait from
a painting by G. HEYWOOD. Dedicated by permission to the Lord
Bishop of Oxford. *Second Edition.* Small 8vo. 6s.
A few Large Paper Edition may still be had, price 10s. 6d.

MAIN's (REV. R.) Twelve Sermons preached on Various
Occasions at the Church of St. Mary, Greenwich. By R. MAIN, M.A.,
Radcliffe Observer at Oxford. 12mo. 5s.

LIST OF WORKS PUBLISHED BY

MASKEW's (Rev. T. R.) Annotations on the Acts of the Apostles. Original and selected. Designed principally for the use of Candidates for the Ordinary B.A. Degree, Students for Holy Orders, &c., with College and Senate-House Examination Papers. By the Rev. T. R. Maskew, M.A. *Second Edition, enlarged.* 12mo. 6s.

MILL's (Rev. Dr.) Observations on the attempted Application of Pantheistic Principles to the Theory and Historic Criticism of the Gospels. By W. H. Mill, D.D., late Regius Professor of Hebrew in the University of Cambridge. *Second Edition, with the Author's latest notes and additions.* Edited by his Son-in-Law, the Rev. B. Webb, M.A. 8vo. 14s.

———— Lectures on the Catechism. Delivered in the Parish Church of Brasted, in the Diocese of Canterbury. By W. H. Mill, D.D. Edited by the Rev. B. Webb, M.A. Fcap. 8vo. 6s. 6d.

———— Sermons preached in Lent 1845, and on several former occasions, before the University of Cambridge. By W. H. Mill, D.D. 8vo. 12s.

———— Four Sermons preached before the University on the Fifth of November and the three Sundays preceding Advent, in the year 1848. By W. H. Mill, D.D. 8vo. 5s. 6d.

———— An Analysis of the Exposition of the Creed, written by the Right Reverend Father in God, J. PEARSON, D.D., late Lord Bishop of Chester. Compiled, with some additional matter occasionally interspersed, for the use of Students of Bishop's College, Calcutta. By W. H. Mill, D.D. *Third Edition, revised and corrected.* 8vo. 5s.

MISSION LIFE among the Zulu-Kafirs. Memorials of Henrietta, Wife of the Rev. R. Robertson. Compiled chiefly from Letters and Journals written to the late Bishop Mackenzie and his Sisters. Edited by Anne Mackenzie. [*In the Press.*

MORRIS' (John) Words of Comfort for the Wayfarer, the Weary, the Sick, and the Aged. Taken from the Writings of the Wise and Good. With an Introduction, by John Morris. Demy 8vo.

NEALE's (Rev. J. M.) Seatonian Poems. By the Rev. J. M. Neale, M.A., late Scholar of Trinity College. Fcap. 8vo. 6s.

NEWTON (Sir Isaac) and Professor Cotes, Correspondence of, including Letters of other Eminent Men, now first published from the originals in the Library of Trinity College, Cambridge; together with an Appendix containing other unpublished Letters and Papers by Newton; with Notes, Synoptical View of the Philosopher's Life, and a variety of details illustrative of his history. Edited by the Rev. J. Edleston, M.A., Fellow of Trinity College, Cambridge. 8vo. 15s.

PEARSON's (Rev. J. B.) The Divine Personality, being a Consideration of the Arguments to prove that the Author of Nature is a Being endowed with liberty and choice. The Burney Prize Essay for 1864. By J. B. Pearson, B.A., Scholar of St. John's College, and Curate of St. Michael's Church, Cambridge. 8vo. 1s. 6d.

PIEROTTI's (Ermete) Jerusalem Explored: being a Description of the Ancient and Modern City, with upwards of One Hundred Illustrations, consisting of Views, Ground-plans, and Sections. By Ermete Pierotti, Doctor of Mathematics, Captain of the Corps of Engineers in the army of Sardinia, Architect-Engineer to his Excellency Soorayn Pasha of Jerusalem, and Architect of the Holy Land. 2 vols. Imperial 4to. 5l. 5s.

——— The Customs and Traditions of Palestine Compared with the Bible, from Observations made during a Residence of Eight Years. By Dr. Ermete Pierotti, Author of "Jerusalem Explored." 8vo. 5s.

PHILLIPS' (Rev. Geo.) Short Sermons on Old Testament Messianic Texts, preached in the Chapel of Queens' College, Cambridge. By the Rev. Geo. Phillips, D.D., President of the College. 8vo. 5s.

PSALTER (The) or Psalms of David in English Verse. With Preface and Notes. By a Member of the University of Cambridge. Dedicated by permission to the Lord Bishop of Ely, and the Reverend the Professors of Divinity in that University. 5s.

ROMILLY's (Rev. J.) Graduati Cantabrigienses: sive Catalogus exhibens nomina eorum quos ab anno academico admissionum 1760 usque ad decimum diem Oct. 1856. Gradu quocunque ornavit Academia Cantabrigiensis, e libris subscriptionum desumptus. Cura J. Romilly, A.M., Coll. SS. Trin. Socii atque Academicæ Registrarii. 8vo. 10s.

SCHOLEFIELD's (Prof.) Hints for some Improvements in the Authorised Version of the New Testament. By the late J. Scholefield, M.A. Fourth Edition. Fcap. 8vo. 4s.

SCRIVENER's (F. H.) Plain Introduction to the Criticism of of the New Testament. With 40 Facsimiles from Ancient Manuscripts. For the Use of Biblical Students. By F. H. Scrivener, M.A., Trinity College, Cambridge. 8vo. 15s.

——— Codex Bezæ Cantabrigiensis. Edited, with Prolegomena, Notes, and Facsimiles. By F. H. Scrivener, M.A. 4to. 26s.

SCRIVENER's (F. H.) A Full Collation of the Codex Sinaiticus with the Received Text of the New Testament; to which is prefixed a Critical Introduction. By F. H. Scrivener, M.A. Fcap. 8vo. 5s.

> "Mr. Scrivener has now placed the results of Tischendorf's discovery within the reach of all in a charming little volume, which ought to form a companion to the Greek Testament in the Library of every Biblical student."—*Reader.*

—————— An Exact Transcript of the CODEX AUGIENSIS, Græco-Latin Manuscript in Uncial Letters of S. Paul's Epistles, preserved in the Library of Trinity College, Cambridge. To which is added a Full Collation of Fifty Manuscripts containing various portions of the Greek New Testament deposited in English Libraries: with a Full Critical Introduction. By F. H. Scrivener, M.A. Royal 8vo. 26s.

The CRITICAL INTRODUCTION *is issued separately, price* 5s.

—————— Novum Testamentum Græcum, Textus Stephanici, 1550. Accedunt variæ lectiones editionum Bezæ, Elzeviri, Lachmanni, Tischendorfii, et Tregellesii. Curante F. H. Scrivener, M.A. 16mo. 4s. 6d.

An Edition on Writing-paper for Notes. 4to. *half-bound.* 12s.

SELWYN's (PROFESSOR) Excerpta ex reliquiis Versionum, Aquilæ, Symmachi, Theodotionis, a Montefalconio aliisque collecta. Gazmia. Edidit GUL. SELWYN, S.T.B. 8vo. 1s.

—————— Notæ Criticæ in Versionem Septuagintaviralem. Exodus, Cap. I.—XXIV. Curante GUL. SELWYN, S.T.B. 8vo. 3s. 6d.

—————— Notæ Criticæ in Versionem Septuagintaviralem. Liber NUMERORUM. Curante GUL. SELWYN, S.T.B. 8vo. 4s. 6d.

—————— Notæ Criticæ in Versionem Septuagintaviralem. Liber DEUTERONOMII. Curante GUL. SELWYN, S.T.B. 8vo. 4s. 6d.

—————— Origenis Contra Celsum. Liber I. Curante GUL. SELWYN, S.T.B. 8vo. 3s. 6d.

—————— Testimonia Patrum in Veteres Interpretes, Septuaginta, Aquilam, Symmachum, Theodotionem, a Montefalconio aliisque collecta paucis Additis. Edidit GUL. SELWYN, S.T.B. 8vo. 6d.

—————— Horæ Hebraicæ. Critical and Expository Observations on the Prophecy of Messiah in Isaiah, Chapter IX., and on other Passages of Holy Scripture. By W. SELWYN, D.D., Lady Margaret's Reader in Theology. *Revised Edition, with Continuation.* 8s.

SINKER's (Rev. R.) The Characteristic Differences between the Books of the New Testament and the immediately preceding Jewish and the immediately succeeding Christian Literature, considered as an evidence of the Divine Authority of the New Testament. Being the Hulsean Prize Essay for 1864. By the Rev. R. Sinker, Trinity College. Small 8vo. 4s. 6d.

STUDENT'S GUIDE (The) to the University of Cambridge. Fcap. 8vo. 4s. 6d.

 Contents: Introduction, by J. R. Seeley, M.A.—On University Expenses, by the Rev. H. Latham, M.A.—On the Choice of a College, by J. R. Seeley, M.A.—On the Course of Reading for the Classical Tripos, by the Rev. R. Burn, M.A.—On the Course of Reading for the Mathematical Tripos, by the Rev. W. M. Campion, B.D.—On the Course of Reading for the Moral Sciences Tripos, by the Rev. J. B. Mayor, M.A.—On the Course of Reading for the Natural Sciences Tripos, by Professor Liveing, M.A.—On Law Studies and Law Degrees, by Professor J. T. Abdy, LL.D.—Medical Study and Degrees, by G. M. Humphry, M.D.—On Theological Examinations, by Professor E. Harold Browne, B.D.—Examinations for the Civil Service of India, by the Rev. H. Latham, M.A.—Local Examinations of the University, by H. J. Roby, M.A.—Diplomatic Service.—Detailed Account of the several Colleges.

TERTULLIANI Liber Apologeticus. The Apology of Tertullian. With English Notes and a Preface, intended as an Introduction to the Study of Patristical and Ecclesiastical Latinity. By H. A. Woodham, LL.D. *Second Edition.* 8vo. 8s. 6d.

TODD's (Rev. J. F.) The Apostle Paul and the Christian Church at Philippi. An Exposition Critical and Practical of the Sixteenth Chapter of the Acts of the Apostles and of the Epistles to the Philippians. By the late Rev. J. F. Todd, M.A., Trinity College, Cambridge. 8vo. 9s.

TURTON's (Bishop) The Holy Catholic Doctrine of the Eucharist, considered in reply to Dr. Wiseman's Argument from Scripture. By T. Turton, D.D., late Bishop of Ely. 8vo. 8s. 6d.

VERSES and Translations. By C. S. C. *Third Edition.* Fcap. 8vo. 5s.

WATERLOO. A Lay of Jubilee for June 18, A.D. 1815. *Second Edition.* 3s.

WIESELER's Chronological Synopsis of the Four Gospels. Translated by the Rev. E. Venables, M.A. 8vo. 15s.

WEST's (C. A.) Parish Sermons, according to the order of the Christian Year. By the late C. A. West, B.A. Edited by J. R. West, M.A. 12mo. 6s.

WHEWELL's (Rev. Dr.) Elements of Morality, including Polity. By the Rev. W. Whewell, D.D., Master of Trinity College, Cambridge. Fourth Edition, in 1 vol. 8vo. 15s.

——————— Lectures on the History of Moral Philosophy in England. By the Rev. W. Whewell, D.D. *New and Improved Edition, with additional Lectures.* Crown 8vo. 8s.

The Additional Lectures are printed separately in Octavo for the convenience of those who have purchased the former Edition. 3s. 6d.

——————— Astronomy and General Physics considered with reference to Natural Theology (Bridgewater Treatise). By W. Whewell, D.D. *New Edition*, small 8vo. (Uniform with the Aldine.) 5s.

——————— Sermons preached in the Chapel of Trinity College, Cambridge. By W. Whewell, D.D. 8vo. 10s. 6d.

——————— Butler's Three Sermons on Human Nature, and Dissertation on Virtue. Edited by W. Whewell, D.D. With a Preface and a Syllabus of the Work. *Fourth and Cheaper Edition*. Fcap. 8vo. 2s. 6d.

WILLIS' (Rev. R.) The Architectural History of Glastonbury Abbey. By the Rev. R. Willis, F.R.S., Jacksonian Professor. With Illustrations.

WINFRID, afterwards called Boniface. A.D. 680—755. Fcp. 4to. 2s.

WILLIAMS' (Rowland) Rational Godliness. After the Mind of Christ and the Written Voice of the Church. By Rowland Williams, D.D., Professor of Hebrew at Lampeter. Crn. 8vo. 10s. 6d.

——————— Paraméswara-jnyána-goshthi. A Dialogue of the Knowledge of the Supreme Lord, in which are compared the claims of Christianity and Hinduism, and various questions of Indian Religion and Literature fairly discussed. By Rowland Williams, D.D. 8vo. 12s.

WOLFE's (Rev. A.) Family Prayers and Scripture Calendar. By the Rev. A. Wolfe, M.A., late Fellow and Tutor of Clare College, Cambridge, Rector of Forsham All Saints, Bury St. Edmund's. Fcp. 1s.

WRATISLAW's (A. H.) Notes and Dissertations, principally on Difficulties in the Scriptures of the New Covenant. By A. H. Wratislaw, M.A., Head Master of Bury St. Edmund's School, formerly Fellow of Christ's College, Cambridge. 8vo. 7s. 6d.

CAMBRIDGE.—PRINTED BY JONATHAN PALMER.

www.ingramcontent.com/pod-product-compliance
Lightning Source LLC
Chambersburg PA
CBHW020901160426
43192CB00007B/1020